THE PHYSICAL ENVIRONMENT OF THE FAEROE ISLANDS

MONOGRAPHIAE BIOLOGICAE

VOLUME 46

Editor
J. ILLIES
Schlitz, F.R.G.

Dr W. Junk Publishers The Hague-Boston-London 1982

THE PHYSICAL ENVIRONMENT OF THE FAEROE ISLANDS

edited by
G. K. RUTHERFORD

Dr W. Junk Publishers The Hague-Boston-London 1982

Distributors:

for the United States and Canada

Kluwer Boston, Inc.
190 Old Derby Street
Hingham, MA 02043
USA

for all other countries

Kluwer Academic Publishers Group
Distribution Center
P.O. Box 322
3300 AH Dordrecht
The Netherlands

Library of Congress Cataloging in Publication Data

Main entry under title:

The physical environment of the Faeroe Islands.

(Monographiae biologicae; v. 46)
Bibliography:
Includes index.
1. Natural history — Faroe Islands. 2. Faroe Islands. I. Rutherford, G. K. II. Series.
QP1.P37 vol. 46 [QH165] 574s [508.491'5] 81-13692 AACR2

ISBN-13: 978-94-009-7979-6 e-ISBN-13: 978-94-009-7977-2
DOI: 10.10007/978-94-009-7977-2

Contents

Legend to frontispiece

A. A Regosol with deep organic surface horizons overlying more strongly humified material.
B. Podzol with an incipient Ae horizon.
C. A Podzol showing a thin ferruginous horizon.
D. Brunisolic soil developed in stone free material. A gravel layer at knife overlies marked organic material.
E. Thick alternating layers of peaty material and stratified sands and gravels.

C
D
E

1. Introduction

G. K. Rutherford

The Faeroes were first inhabited by Irish monks in the year 700 A.D., yet they have remained insular and isolated for over 1000 years. In 1948 the Faeroes were still relatively unknown in the Northern European economic community and were regarded as an underdeveloped nation. The main occupation was fishing, as it still is now. No other primary activities existed other than local fishing and subsistence agriculture. There were no factories or industries. Enormous development has taken place since then and this development has created a demand for an inventory of the Faeroe Island natural resources.

Although scientific investigations have been conducted of the Faeroes since 1673, it was not until 1969 that Denmarks Geologiske undersøgelse completed the first detailed geologic study. This report (Rasmussen and Noe-Nygaard 1970) discusses the Faeroe Islands' Plateau Basalts. Prior to this investigation little was known about the geology of this island archipelago. The earlier reports are confined primarily to a descriptive recording of the bedrock and glacial geology (Debes 1673, Forchhammer 1824, Grossman and Lomas 1895), botany (Trevellyan 1835, Ostenfeld 1905), and zoology (Jensen, et al. 1901). Generalized descriptions of the landscape occur briefly under the topic heading of glacial geology (Geike 1880) during this period. Ødun (1923) was first to systematically report the overall character of the Faeroese landscape and his original report has only recently been elaborated upon by Trap (1968).

In order to develop an understanding of the relationships between the variables of the landscapes (geology, topography, geomorphology, soils vegetation and land use), a land classification and soil survey was conducted. By establishing the physical and micromorphological characteristics, chemical properties and classification of the soil materials and landscape, the framework can be useful for the management of the land. This, in turn, enables an understanding of the distribution and pedogenesis of the soils in the Faeroes.

The spatial distribution of the soils with the geology and topography is presented within the genetic, hierarchial framework of the land systems approach. This approach allows the breakdown of the overall landscape into areas of similarity by the concept that individual geomorphic processes produce

Monographiae Biologicae, Vol. 46 ed. by G. K. Rutherford

distinctive landscape features. These features or landforms are described by their morphology, their materials and wherever possible by their origin (Lacate 1969). The ability to predict landforms and soil properties is derived from this understanding of the relationships between individual landform systems (Valentine 1970).

In addition, the spatial relationships of the soils allow for a fuller interpretation of the soils and the factors responsible for their formation. In order to reveal the nature, distribution, and origin of the major soil groups found in the Faeroes, fully documented soil samples were investigated.

In this report, the landscape is examined using land systems: each land system is illustrated and in turn is divided into smaller, more homogeneous areas. The distribution of these latter areas within the land systems and their soils is indicated by vertical cross sections and tabular descriptions.

Specific terms used by the Faeroese to describe the geomorphic features of their landscapes have been referenced. These terms have either been underlined and/or placed in parenthesis. The definitions of these terms occur within the text, but more explicit descriptions are presented in Appendix 1.

Author's address:
Dr G. K. Rutherford
Department of Geography,
Queen's University,
Kingston, Ontario,
Canada.

2. Location and general description

G. K. Rutherford & C. E. B. Taylor

The Faeroe Islands (62°N, 7°W) are located between Iceland and Scotland, about 782 km from Reykjavik and 300 km from the Shetlands (Fig. 2.1). The interplay between the constructive forces of vulcanism in forming the landscape and the destructive forces of water, weathering frost action, glacial and marine erosion has produced a harsh and rugged terrain. The twenty-one islands, skerries and outcroppings occupy a total land mass of 1399 km^2 (139 885 ha). The islands present a series of ridges and hill summits which project above sea level from a broad submarine ridge that separates the Norwegian Sea from the North Atlantic Ocean (Rasmussen 1963) (Fig. 2.2).

The islands are a self governing society within the Danish Kingdom and, in many respects, may be regarded as an independent state. The 'Lagting' is the name of the Faeroese Parliament. The Home Government is vested with total authority over governmental form, municipal affairs, direct and indirect taxation, health, education, social services and transport. The Faeroese Home Rule Law of 1948 distinguishes between special affairs and joint affairs. There are a number of affairs which both the Danish and Home Government are concerned with and the Danish authorities have final word in foreign affairs.

The people are mainly of West Norwegian extraction and have lived in semi-isolation for over a thousand years, developing their own language and cultural traditions. The eighteen islands in the North Atlantic are exposed to an unrelenting attack by wind and weather which together with the prevailing ocean currents produces a cool, humid and windy climate.

Although agriculture has been the predominant way of life through many centuries, the islanders enthusiastically developed a fishing fleet in the early part of this century. The export income is now overwhelmingly derived from fish and fish products. The economic conditions prevailing since the end of the Second World War have enabled the community to establish an enviable standard of living and life style, which compares favourably with other Scandinavian countries.

The total population is approximately 42 000, with 12 000 in the capital Torshavn on the main island of Streymoy. Klaksvik, on Bordoy, has about 5000

Monographiae Biologicae, Vol. 46 ed. by G. K. Rutherford

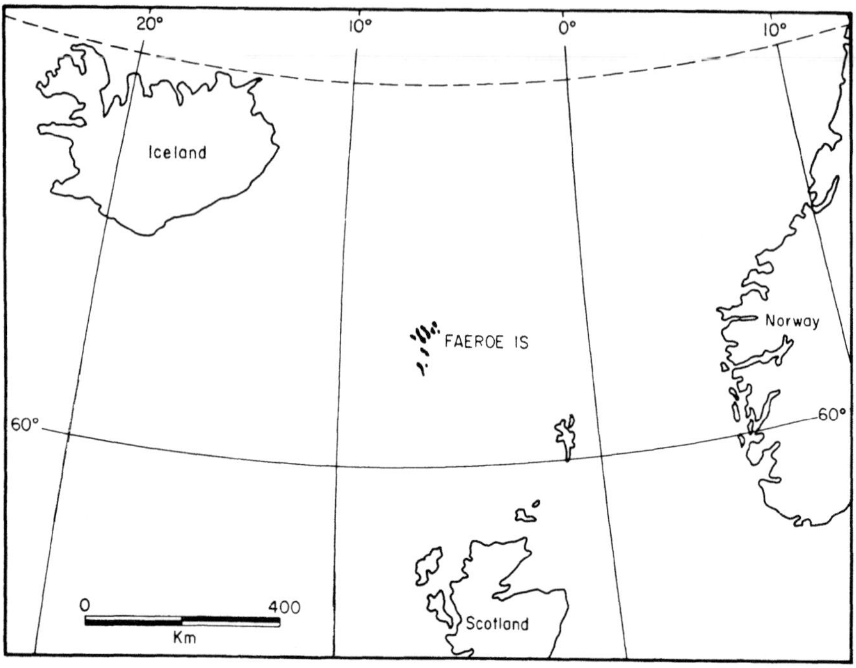

Fig. 2.1. The location of the Faeroe Islands.

people; Tvøroyri and Vagur on the southern island of Suderoy, have about 2000 each. Vestmanna, on Streymoy, ranks fifth in population with 1400. The population is dispersed relatively evenly throughout the islands, as the agricultural community has lived in small hamlets and isolated farms.

Weather permitting, there are frequent air connections with Denmark and Iceland and regular car/passenger sailings to Scotland, Norway and Iceland. During the summer, there is a frequent passenger boat service to Denmark. Although there is a good road network on all the larger islands, the majority of passenger and goods transport is still reliant upon frequent and fast boat service to the many isolated coastal communities.

The Faeroe Islands were formed by successive lava flows extruded from large fissure eruptions in the high lying Wyville-Thompson Ridge (Schrøder 1973) (Plate 2.1). This ridge is an elongated shallow water feature trending SE to NW, and the Faeroes constitute a stepping stone between the Scottish and the Icelandic part of the Brito-Arctic Basalt Province (Noe-Nygaard & Rasmussen 1957). Plateau basalts of this Province are found in northeast Ireland, the Inner Hebrides, Iceland, Jan Mayen, West Greenland and part of East Greenland. The age of volcanic activity has been dated at 50 000 000 to 60 000 000 years B.P., and is consistent with the Lower Eocene age volcanics observed in Scotland and East Greenland (Tarling & Gale 1968).

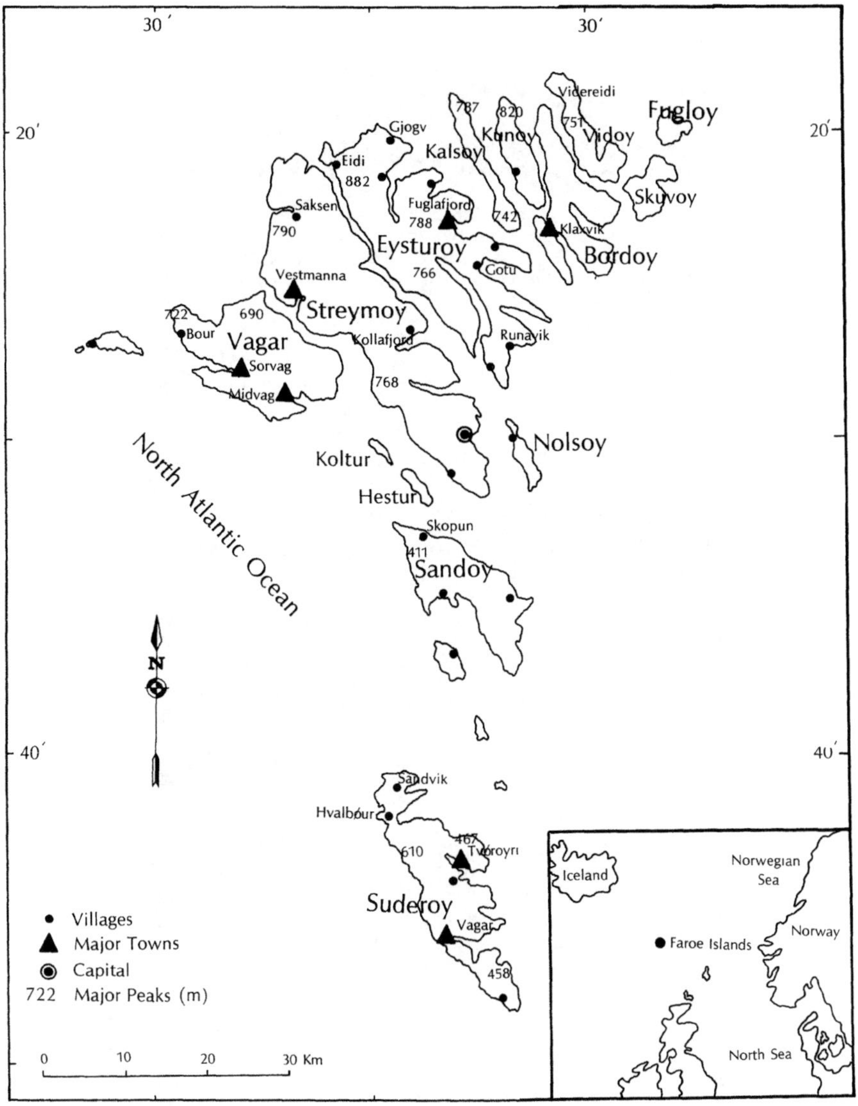

Fig. 2.2. The eighteen islands that comprise the Faeroes.

The lower series of lavas consists of forty to fifty flows, ranging in thickness from 10–30 m. The middle series is approximately 1300 m thick and consists of lava flows ranging in thickness from several centimetres to two metres. The upper series, like the lower series, were extruded in a pulsating fashion, but the flows are less massive and well intercalated with tuffs.

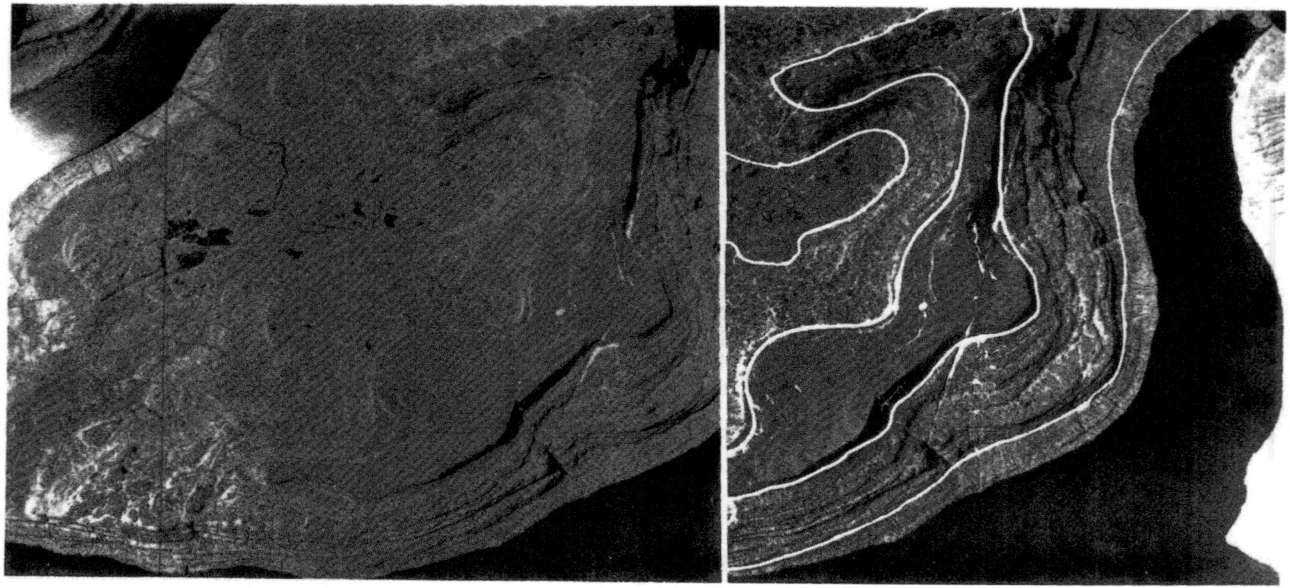

Plate 2.1. Panchromatic aerial photograph taken on May 10, 1970 from an altitude of 12 000 m. (Reproduced from Geodaetisk Institut Danmark.)

Dykes intruded into steeply inclined fracture lines are a common feature of the Faeroese landscapes, and often exhibit columnular structure. More often, the dyke is less resistant than the surrounding terrain and erodes to form a long, steepsided gully (gjogv). Sills are not common in the Faeroes plateau, but several exist as intrusions between the middle and upper series.

The Faeroes have experienced a local ice sheet but the glacial striae are indistinct and radiate from the centre of the island, there is no evidence of ice movement from the north (Geike 1880); the roche moutonée have their smooth slope pointing inland and terminal moraines are non-existent (Killerich 1905).

The polycyclic nature of the Faeroese landscape is revealed in the juxtaposition of features carved by glacial erosion and then modified by nival and fluvial erosion. An example of this landscape is presented in the unique cirque form (botnur) carved in the mountains of the Northern islands (Plate 2.2). When two neighbouring botnur erode into a mountain from opposite sides, the mountain ridge may have been reduced to a narrow saddle or col (skard). A broader and narrower col is termed a halsúr.

The most significant influence of the last glaciation on the Faeroese landscape was the removal of the mass of decomposed material which had accumulated since the Tertiary on the mountain slopes and plateau areas (Killerich 1905, Currie 1960).

Glacial till and boulder clay occur in the Faeroes as localized deposits predominantly on the lee sides of roche mountonées or along the margins of the

Plate 2.2. Unique cirque form carved into the mountains of the northern islands.

more gently sloping valleys, wide fjords and eidis (Geike 1880, Rasmussen 1963) (Plate 2.3). An eidi is an isthmus or low, flat strip of land that separates two fjords, which trends east-west in the Faeroes. Mass wasting, terracettes, solifluction deposits and sorted patterned ground have been described in the Faeroes by numerous authors, but only recently have the processes of solifluction exhibited in the Faeroes received investigation (Lewis & Lass 1965).

Lower limits of widespread contemporary solifluction occur at 300 m in the Shetlands, 600 m in the Northern Highlands of Scotland and at 750 m in the Western Highlands of Scotland (Romans personal communication), whereas in the Faeroes sorted patterned ground and solifluction deposits were observed as low as 200 m a.s.l. Geike (1880) reports that no raised beaches are evident because the Faeroes belong to the region of Mesozoic and Tertiary subsidence that extended southwards to the North Sea. Daly (1934) reported that the Faeroes showed no signs of emergence in response to the removal of a thick ice cap. Recently, Sundt (1976) states that a certain isostatic rebound is evident; field observations of the ancient Viking landing site at Vikar (Vagar), now 10 m out of water and minor strand lines at Leyna (Streymoy) suggest that tectonic uplift is occurring along with a slow isostatic rebound of the land mass.

The rugged topography and mountainous relief of the Faeroes is a result of tectonic upbuilding and subsequent reduction by glacial, periglacial and fluvial erosion. The average elevation of the Faeroes is 300 m a.s.l. The northern islands have the greatest relief: over 1000 peaks exist at 300 m and 10 peaks are greater than 800 m elevation. The summits are within 2 km horizontally of their base and slopes of greater than 40° are not uncommon.

Plate 2.3. View of the gently sloping valley of a wide Faeroese fjord where glacial till often occurs as localized deposits.

The general trend of the Faeroes' major valleys and fjord system is NNW-SSE and has been dictated by the prominent trend of the Wyville-Thompson Ridge (Rasmussen & Noe-Nygaard 1970). To construct a line from the most western island (Mykines) to the most eastern island (Fugloy) would result in a line perpendicular to the fjords. This line passes through numerous eroded cols

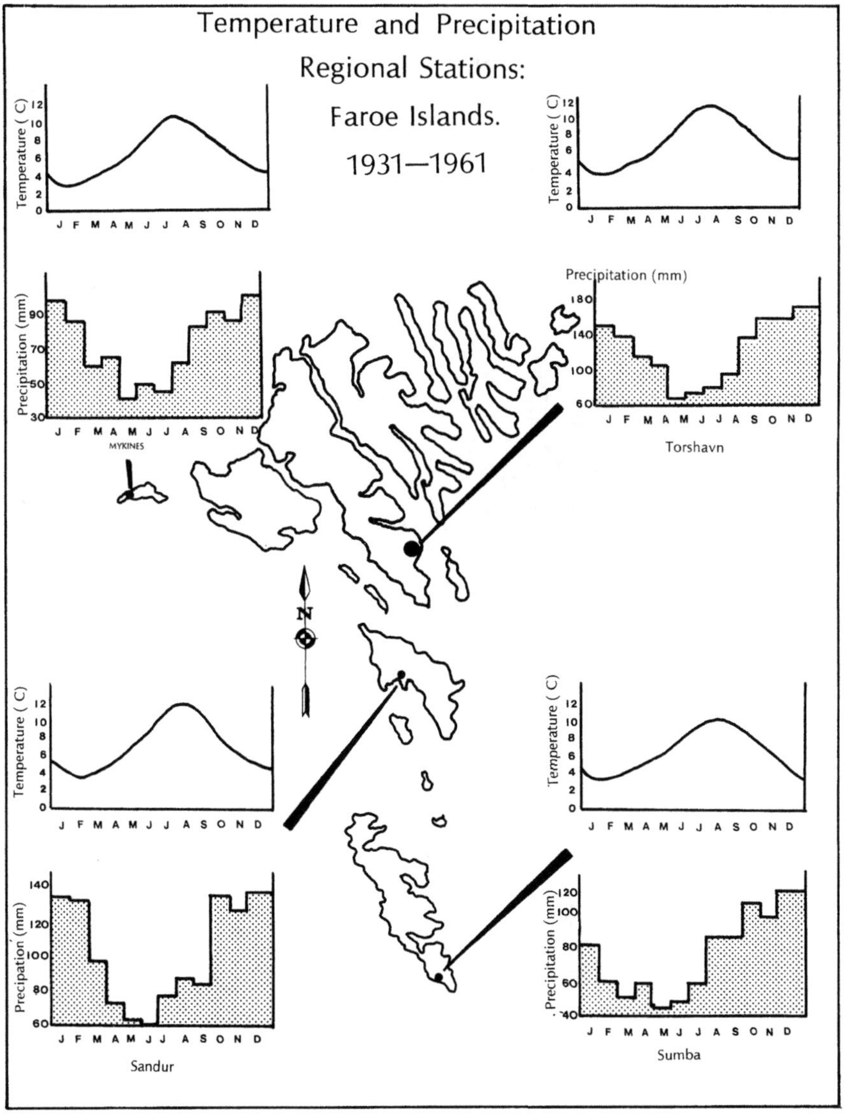

Fig. 2.3. Temperature and precipitation data.

and saddles, marking the line of ancient icesheds and the present watersheds (Grossman & Lomas 1896).

Abrupt cliffs (Kunoyarnakkur, 820 m) and sea stacks (Bugvin, 210 m) dominate the west coast, and in general the watershed remains very close to the western coast. On Streymoy, Snubbing rises to 623 m within a horizontal distance of 180 m from the west coast (angle of 75°) and then descends to the east coast at an average of 30°.

The climate of the Faeroes is characterized as cool, wet, windy, and overcast (Koppem: Cfc). The Faeroes are situated at the confluence of two different water masses. One water mass is of polar origin (East Icelandic Polar Current) and the other is of tropical origin (North Atlantic Drift). These water masses are of

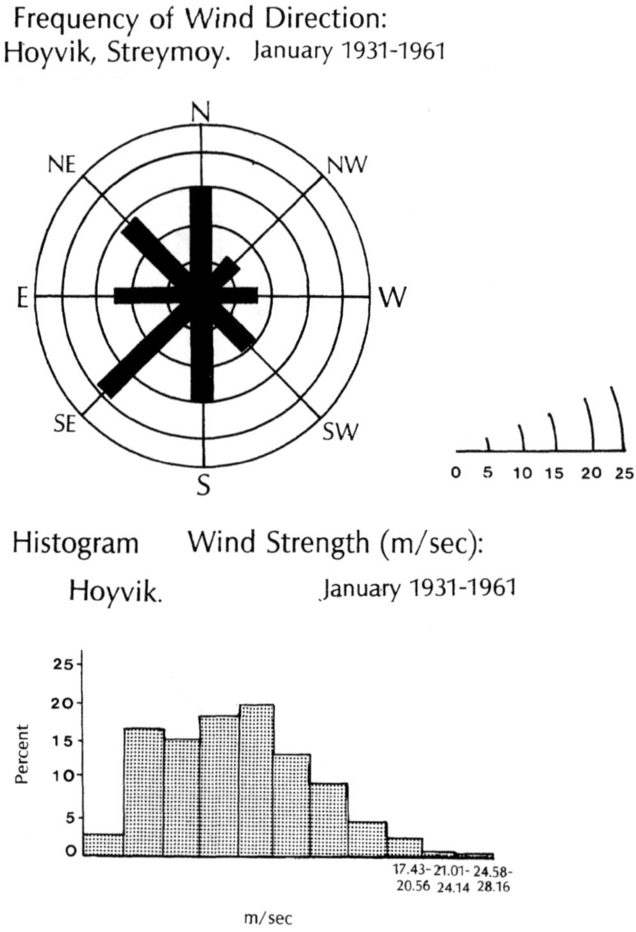

Fig. 2.4a. Wind strength and direction (January); Hoyvik, Streymoy.

primary importance, and it is the hydrographic conditions that modify and maintain the small range in annual temperatures, short summer season and winter maximum of precipitation experienced in the Faeroes.

The average annual temperature and precipitation data recorded at four stations in the Faeroes are presented in Fig. 2.3. Slight regional differences in the data are apparent, particularly the slightly warmer conditions recorded in Suderoy. The climatic data recorded at Torshavn represent the general climatic conditions experienced in the Faeroes.

Frost occurs as often as 68 days a year during the months of December through March. The air temperature rarely remains below 0°C for a full 24 h (Lysgaard 1969). Precipitation occurs over 281 days a year and averages

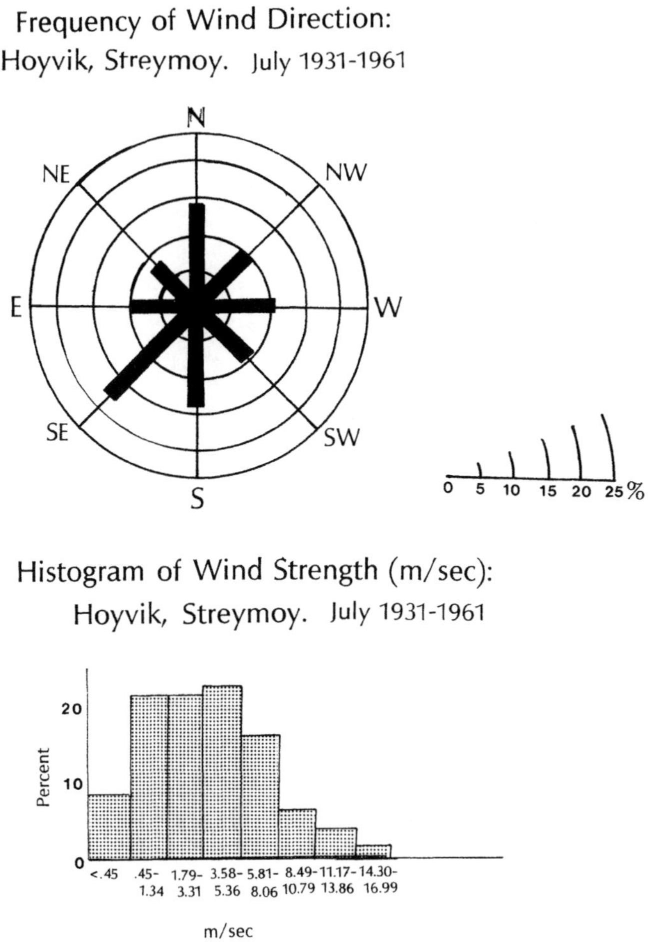

Fig. 2.4b. Wind strength and direction (July); Hoyvik, Streymoy.

1461 mm annually, although local variations are noticed. Over 60 % of the rainfall occurs from October to January. Although snow falls on the average of 44 days a year, it very rarely remains for more than a day. The relative humidity varies little and the yearly average is 83 %. Skies are generally cloudy as is indicated by an average recorded overcast value of 78 %. Fog occurs frequently, and although it is commonly localised over a region, it seldom remains for more than 8 h.

The wind is constantly shifting due to fluctuations in atmospheric pressure. Thirty percent of the winds are from the SW and W and 26 % are accounted for from the N and NE. The strength of the wind is of particular interest, occuring frequently at velocities greater than 8.5 m/s during the winter months (Figs. 2.4a, 2.4b).

Authors' addresses:
Dr G. K. Rutherford
Dept. of Geography,
Queen's University,
Kingston, Ontario,
Canada.

Dr C. E. B. Taylor
Ontario Hydro,
Toronto, Ontario,
Canada.

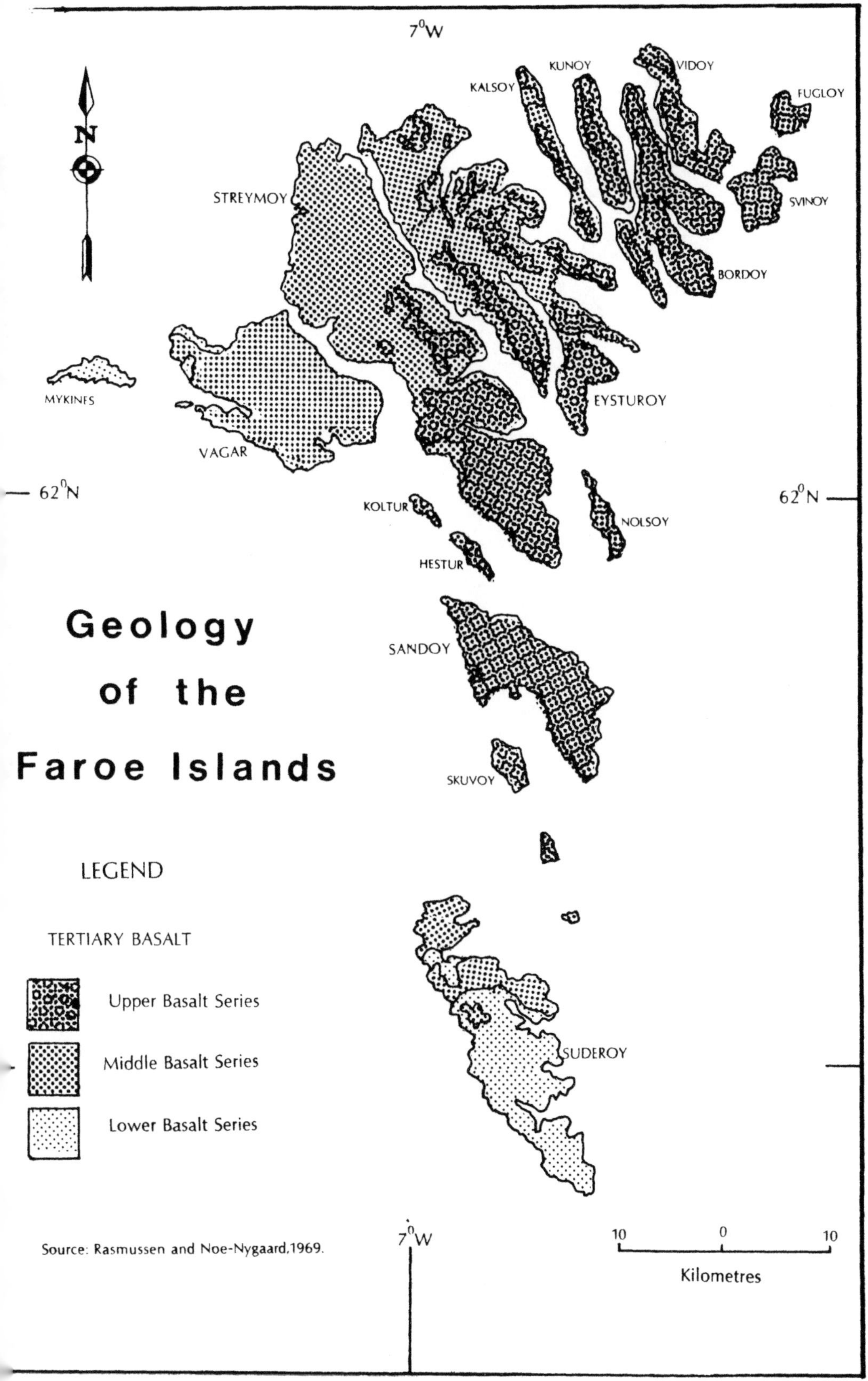
7°W
N
KUNOY
VIDOY
KALSOY
FUGLOY
STREYMOY
SVINOY
BORDOY
MYKINES
EYSTUROY
VAGAR
62°N
62°N
KOLTUR
NOLSOY
HESTUR
Geology of the Faroe Islands
SANDOY
SKUVOY
LEGEND
TERTIARY BASALT
Upper Basalt Series
Middle Basalt Series
Lower Basalt Series
SUDEROY
Source: Rasmussen and Noe-Nygaard, 1969.
7°W
10
0
10
Kilometres

3. The Faeroe Islands: geology

J. Rasmussen

The Faeroe plateau

The Faeroes compose an area of 1399 km^2; the highest point, Slættaratindur, is 882 m above sea level. They are built up by basaltic lava flows with interbedded pyroclastic material of lower Tertiary age (50–60 million years b.p.). The total thickness of visible basalt is ca. 3000 m; these are divided into the lower, the middle and the upper basalt series. Between the lower and the overlying series, there was a longer period of quiescence, which is represented by a coal and clay sequence of approximately 10 m.

The archipelago's most marked morphological characteristic is the NW–SE direction of the fiords; a direction which corresponds to the direction of the assumed volcanic fissures. It also corresponds with the direction of the Iceland-Faeroe ridge, the Wyville Thomson ridge and that of the channel between Faeroe Bank and the Faeroes.

The Faeroe plateau is bounded to the northwest by the Faeroe-Iceland ridge. To the east and west, the Faeroe plateau falls steeply off to the deep ocean trenches. It borders at the southeast on the Faeroe-Shetland channel and to the southwest by the Faeroe Bank channel which runs to the southeast out into the Faeroe-Shetland channel.

The Faeroe plateau constitutes the northeastern part of the Faeroe-Rockall plateau which must be considered as a continental fragment, a microcontinent, separated from the northwest European plate due to the ocean floor spreading during the opening of the North Atlantic Ocean.

The North Atlantic Basalt Area

From a geological point of view, the Faeroe Islands belongs to the North Atlantic Basalt Area or the Brito-Artic Basalt Province. Parts of this area also occur in northeast Ireland, the west coast of Scotland, in SE-NW Iceland, on Jan Mayen, in parts of east Greenland, west Greenland and in Baffin Island (Fig. 3.1).

Monographiae Biologicae, Vol. 46 ed. by G. K. Rutherford

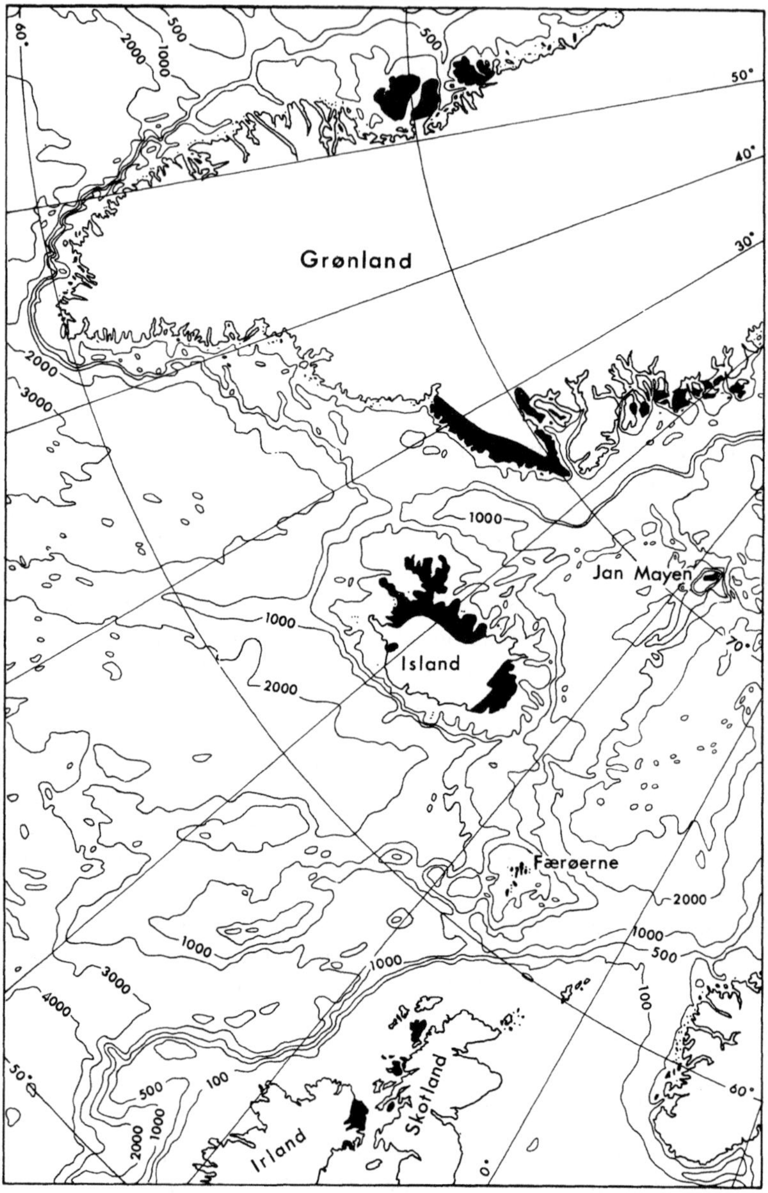

Fig. 3.1. The North Atlantic Basalt province (depth in metres) of which the Faeroes are a member.

Common for all of these individual areas is that they consist of extensive basalt lava flows which have the character of a plateau, leading to the concept of plateau basalts. The term flood basalts is now preferred, referring to the mobility of the lava. The lava pile is composed of individual lava flows, the majority of which emerged through fissures in the earth's crust.

Plateau basalts or flood-basalts of similar composition and structure as the North Atlantic Basalt area can be mentioned: The Deccan basalts in India, the Columbia-Snake River plateau in the United States and the basaltic areas in South America, South Africa and central Siberia. Most of them belong to the earliest part of the Tertiary.

The stepwise developed morphology of plateau-basalts have since early times given rise to the word 'trap' and is due to the different effects of erosion on the hard basalt and the less resistant interbedded layers.

The geological evolution of the Faeroes

The geologic history of the Faeroes can be divided in two main periods, the period of active volcanism when the plateau itself was built-up and the period of destruction when forces of erosion and weathering have given the island the landscape morphology it has today.

The basalt plateau which, as earlier mentioned, is about 3000 m thick, is divided into six series or steps which in ascending stratigraphical order express the progress of volcanism and the relative age of the strata.

6 Intrusive rocks
5 The upper basalt series
4 The middle basalt series
3 The tuff-agglomerates zone
2 The coal-bearing sequence
1 The lower basalt series

From now on, we will attempt to describe the different stages in the building of the plateau as an ideal vertical section through the lava pile, but due to the fact that the general dip is easterly, laterally displaced from the west to the east (Fig. 3.2).

The lower basalt series. The volcanic activity which built up the lower basalt series, was rhythmic in that each eruption has been followed by a period of dormancy.

The volcanic eruptions manifest themselves mainly by the production of lava and in a minor degree volcanic bombs, lapilli, and ashes. The erosion of lava surfaces and deposition of sediments, now and then containing small amounts of coal representing the remains of the vegetation, took place during the period of dormancy.

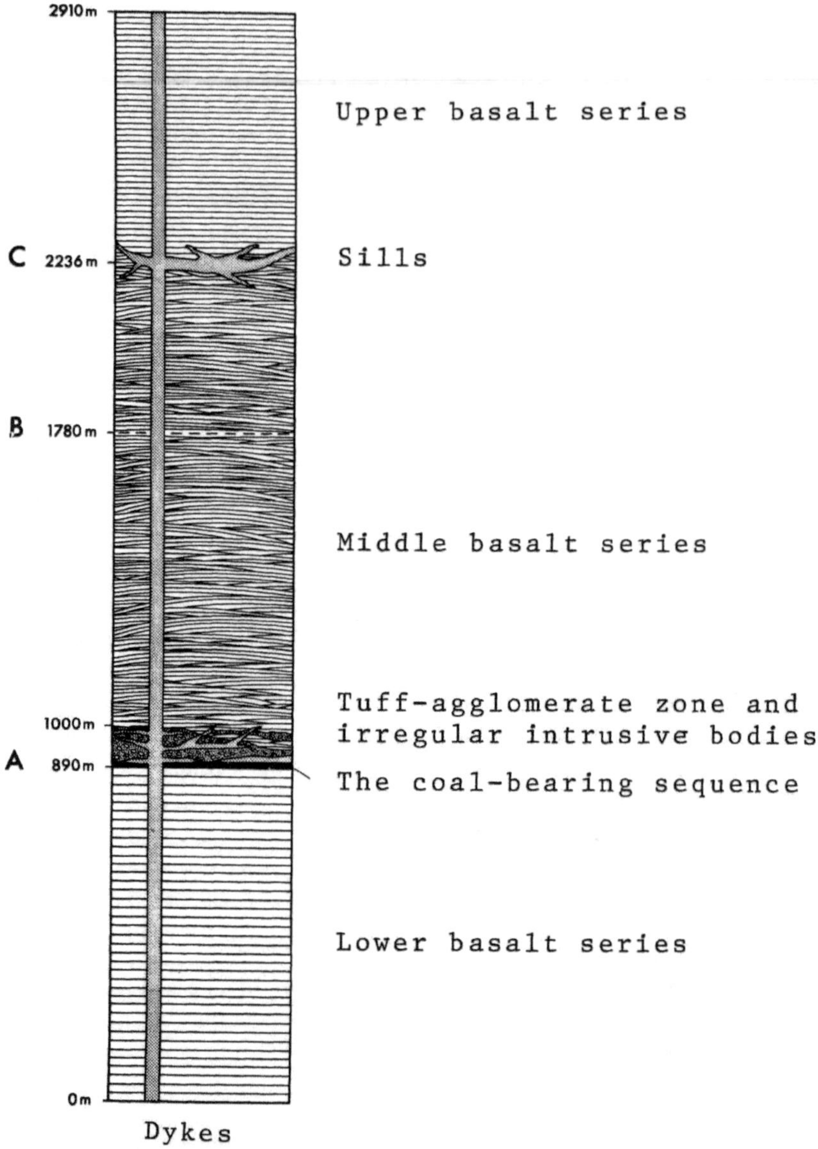

Fig. 3.2. Schematic section through the lava sequence of the Faeroe Islands.

It is considered that the volcanic acitvity has been connected to kilometre long fissures, having the same direction as the Faeroese fiord system, the Wyville Thomson ridge and the Faeroe-Iceland ridge, and that pyroclastic rocks belonging to the tuff-agglomerate zone cover the fissures and thin immediate surroundings.

The lower basal series occurs on Suðuroy, on Mykines, Gáshólmur, Tindhólmur and on the western part of Vágar. On the southern end of Suðuroy, on Mykines, and on Gáshólmur, the lower basalt series dominate whereas on the northern part of Suðuroy, on Tindhólmur and on Vágar, it is covered by younger strata (Plate 3.1).

The total thickness of the lower basalt series is ca. 900 m. The individual basalt flows are commonly 10–30 m thick; thinner flows occur just as there are flows of more than 50 m, the average thickness is about 20 m. The rock is rather uniform; a compact, hard, bluish basalt commonly without phenocryst. The surfaces of the flows are scoriaceous, vesicular and reddish.

The basalts in the lower series often form columns which are a result of the contraction during the cooling of the lava flows. Under ideal cooling conditions, the columns are regularly six-sided, but most commonly they are regularly six- five- four-sided (Plate 3.2). The columns are usually vertical, perpendicular to the cooling walls but can often be fan-shaped or even twisted into bizarre forms.

The interbasaltic sediments are essentially limited to tuffaceous and clay sediments, shales, basaltic sandstone and conglomerates. There are sporadic occurances of coal in these sediments. The tuff beds, which are hardened volcanic ash, show considerable variation in colour. Though usually brownish or reddish and occasionally bright red, they can also be yellow, green or grey-black.

Plate 3.1. The lower series-coal bearing sequence — middle basalt series, as seen on Suðuroy, illustrates the evolution of the basalt plateau.

Plate 3.2. The basalt columns of the lower series are usually vertical or perpendicular to the cooling walls. (Trongisvágur, Suðuroy).

The coal-bearing sequence. After the formation of the lower basalt series, there was a long period without volcanic activity, during which the lava surfaces were eroded and the coal-bearing sediment deposited. The coal-bearing sequence is deposited on a eroded and subaerially weathered uneven surface.

The long period of dormancy is marked not only by the presence of the coal-bearing sequence but also by a tensional adjustment in the plateau. On Suðuroy, there are faults with displacements of more than 10 m, which is older than the middle basalt series and on Vágar, a clear discordance may be seen between the lower and middle basalt series.

The coal-bearing sequence occurs on Suðuroy, Tindhólmur and Vágar. On Suðuroy, the coal-bearing sequence covers an area of ca. 23 km^2 (Fig. 3.3). On Tindhólmur-Vágar the coal has little practical importance. On Suðuroy, the coal-bearing sequence is about 10 m thick. The succession from the bottom upwards is (1) whitish grey underclay, (2) a lower coal seam, (3) dark shale with thin streaks of coal, (4) upper coal seam, (5) clay beds (roof-clay) with small amounts of coal. Fluviatile conglomerates may sometimes take the place of the roofclay.

The thickness of the coal varies from place to place but averages approx-

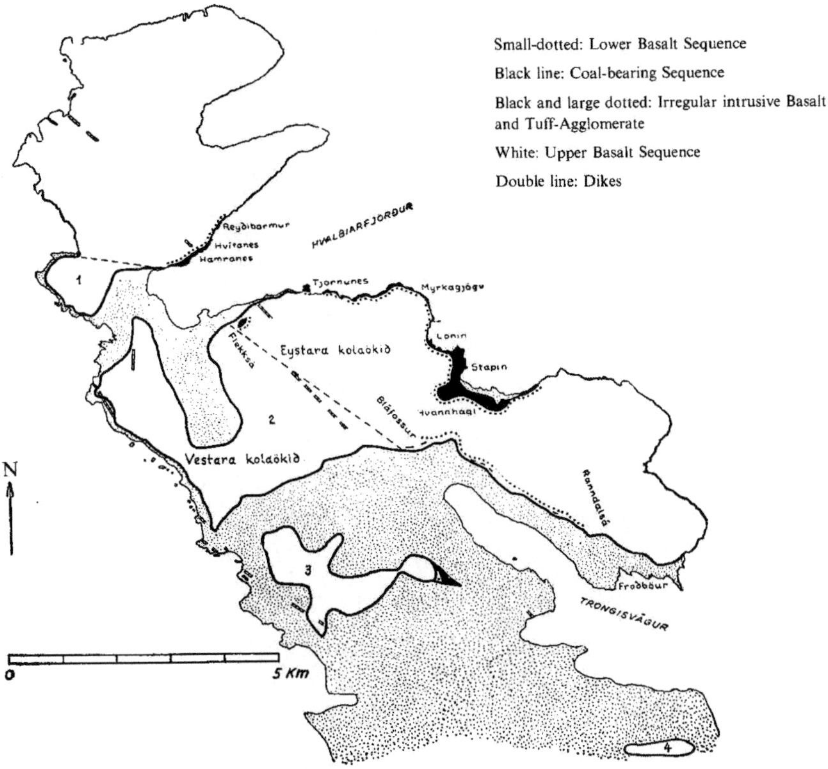

Fig. 3.3. Geological map of the northern part of Suðuroy, distinguishing the coal bearing sequence.

imately 0.75 m in the western part of the sequence. The roof-clay is variable with respect to colour and quality, comprising by far the largest part of the sequence. The coal-bearing sequence is seen in the steep, western mountain walls between the lower and middle basalt series; it dips towards the northeast and reaches sea level on the northeastern side of the island (Plate 3.3). The coal thickness is greatest in the southwest where the sequence is highest, 45 a.s.l., and thins towards the north and east.

The coal is allochthonous, since the vegetation which formed the source material for the coal formation did not grow at the site, but was brought in from surrounding areas. Periodically fragments of more or less carbonized stems and roots are pulled out from the coal mines. These are packed together one on top of the other and never on stumps; something which also indicates the coal is allochthonous.

The macrofossils do not permit an age determination of the coal seams, the few leaf impressions all belong to Metasequoia occidentalis. In later years, however, microfossils such as spores and pollen indicate an Eocean age.

Plate 3.3. The lower basalt series-coal bearing sequence-middle basalt series in the steep mountain walls of West Suðuroy.

The Faeroese coal is very limited and has been used periodically as far back as the 1600's. Since then, coal has regularly been used by either public initiative or for local, private use.

Coal mining was of significant importance during the two world wars, 1914–18 and 1939–45: in the years after the Second World War, most of the domestic fuel was provided by Faeroese coal but recently, coal mining has been of only modest importance. In 1953 the coal resources were estimated to only 12 million tons, some of which have since been mined. The Faeroese coal is intermediate between lignite and bituminous coal. Two different types are distinguished: bright coal (vitrain) and dull coal (durain). The bright coals are shiny, clean and very pure with a calorific value of ca. 6000 Kcal/kg and an ash content of under 5 %. The calorific value of the dull coal is usually 4/5000 Kcal/kg with an ash content up to and exceeding 20 %.

The tuff-agglomerate zone. After a rather lengthy period of quiescence, with deposition of the coal-bearing beds, the volcanic activity started again with a highly explosive initial phase. During this time, lava production was minimal and mainly loose, eruptive products (pyroclastic rochs) such as volcanic bombs, lapilli and ash, were erupted. Hardening of the ash formed tuff, while the coarse material was indurated to agglomerate.

During the explosive phase, the coal-bearing sequence was shattered and the eruptive products were deposited above the coal-bearing beds in the vicinity of the eruption sites. The tuff-agglomerate zone occurs in elongated NW-SE trending areas on the north side of Suðuroy and on the northwest side of Vágar and Tindhólmur; the only two possible observation sites of this zone where the boundary between the lower basalt series, the coal-bearing sequence and the middle basalt series are observed above sea level.

On the northeastern side of Suðuroy, the tuff-agglomerate zone occurs in an area of 10 km × 2 km. It extends from the north side of Hvalbiarfjørður to the northern side of Trongisvágur. On the north side of Hvalbiarfjørður, a remarkable steep, red cliff can be seen where approximately the entire coastal profile, up to 30–50 m in height, is composed of tuff and agglomerates. On the south side of Hvalbiarfjørður and the north side of Trongisvágur, it appears in gullies and rivers courses.

On Vágar, the tuff-agglomerate zone forms the coast around Gásadalur, also occurring in gullies and river courses for 7 km towards the island's west point. On Tindhólmur, it is clearly distinguished in the vertical cliff on the south side of the island between the lower basalt series in the coast and the middle basalt series above (Plate 3.4).

Plate 3.4. The tuff-agglomerate zone occurs above the lower basalt series (at the coast) and is overlain by the middle basalt series at Tindhólmur.

The tuff-agglomerate zone overlies the coal-bearing sequence and is overlain by the middle basalt series. It is intruded by irregular basaltic intrusions. As the tuff-agglomerate zone is an accumulation of a very heterogeneous material, the thickness will vary from place to place, between a few metres to more than 50 m; the average being 20–30 m.

The middle basalt series. As an immediate continuation of the highly explosive initial phase which gave rise to the tuff-agglomerate zone, a lava phase occured with very little explosive activity. The volcanism was continuous, the lava flows were poured out without major pauses. Alternation between tuff-agglomerates and thin basalt flows of the middle basalt series is not uncommon at the boundary between the tuff-agglomerate zone and the middle basalt series.

The middle basalt series is up to 1350 m thick and dominates the northern part of Suðuroy, Vágar, the northern part of Streymoy and the northwestern part of Eysturoy. In other places it is overlain by the upper basalt series.

The thickness of the individual lava flows in the middle basalt series are considerably less than in the other two series, usually from a fraction of a metre to one-two metres; they often form smaller units up to 20 m or more. In the upper part of the series the flow thickness increases somewhat (Plate 3.5).

The lava flows in the middle basalt series are commonly separated by highly vesicular zones, tuff are uncommon and thin in the lower part of this series,

Plate 3.5. The middle basalt series (on Eysturoy) is often composed of smaller units up to 20 metres or more.

though more common in the higher parts. If one considers an individual lava flow, one often finds a zone with more or less vertical pipes at the base; above this is a compact solid middle zone and on the top, a highly vesicular zone where gases have been liberated as a result of the reduced pressure in the lava at the surface. Such pipes and pores are often filled with secondary minerals such as quartz, opals, zeolites and calcite.

Ropy lava surfaces are common; they are formed during the cooling of the surface and show the local flow direction. Flow structures, appear as parallel stripes with crystals and pores oriented in the flow direction. These are sometimes so prominent that they give rise to a lamellar jointing.

The rock composition is considerably more variable in the middle basalt series than in the lower basalt series. There are both fine textured basalts, feldspar basalts and olivine basalts.

In the exposed vertical cliff walls of the archipelago's north and west coasts, the middle series is much more compact than the other two series. Though separated by vesicular zones, they otherwise appear, particularly lowest in the series on the north side of Suðuroy and the west side of Vágar, as an indurated mass in contrast to the other two series, where basalt flows and a tuff layer alternate. Steep mountain walls show clear striping, a result of the wind erosion of the vesicular zones.

Remnants of the vents, which must be considered as feeding channels for the middle basalt series, are seen along fiords and sounds. They vary from less than 50 m up to 1.5 km, where observations seem to show that they must have had an elongated or elliptical form. They are seen in vertical coast sections between the islands as agglomerates, bounded to the sides by regular basalt flows.

The upper basalt series. After a further period of quiescence of unknown duration, during which erosion and weathering of the lava surface took place, the last period of eruption commenced. The surface, on which the first lava flows were poured, was uneven and sloping. Local erosion has left more resistant parts which, in their time, protruded over the surface and were overflowed by a new lava stream.

The volcanic activity which was continuous during the formation of the middle basalt series, now had a rhythmic character again. The volcanic activity in the upper basalt series moved from east to west, but during the building up of the middle basalt series it flowed from west to east.

The upper basalt series, like the lower basalt series, is built up of varying basalt and tuff-clay beds. The average thickness of the lava flows is less than in the lower basalt series; usually between 5 and 10 m (Plate 3.6). In some places, particularly on the northern islands, there are thin lava flows separated by vesicular zones. Columnar structure is not very common and poorly developed, as compared to the lower basalt series. On the other hand, fluidal structures are often well developed, particularly in the fine textured, compact basalts. The

Plate 3.6. The average thickness of lava flows in the upper basalt series is less than the lower basalt series.

rock types are the same as in the middle basalt series, olivine basalt dominates.

Apart from the usual tuff beds, in which there are a few poor examples of stem and leaf impressions, tuffclay depositions occur of considerable size.

The upper basalt series, approximately 675 m thick, does not occur on Suðuroy, Mykines, Gáshólmur, Tindhólmur and Vágar. It is also absent from the northern part of Streymoy and the northwest part of Eysturoy. On the southerly part of Streymoy, Eysturoy and on all the other islands, it overlies the middle basalt series. It dominates the archipelago's east coast.

Intrusives. When the extrusive volcanism was over and the basalt plateau had been formed, the final volcanic phase — the intrusion of the basaltic magma — took place. While the magma during the building up of the plateau rose up to the surface it now penetrated the zones of weakness in the plateau as intrusives. The zones of weakness through which the lava flowed, were partially vertical or almost vertical fissures in the plateau, caused by settling in the underlying support and partially the border zones between the three basalt series. The intrusive rocks occur in three different forms: dykes, irregular intrusives and sills.

Dykes are formed from magma penetrating vertical or inclined fissures in the plateau. They form elongated plate-shaped bodies, which transect the normal basalt flows. In the terrain they are usually straight, but can be curved or gently sinuous. The dykes have a more or less pronounced columnar structure, the columns are oriented at right angles to the walls. The rock type found in the dykes is the same as that of the other three basalt series, but they are more compact. One might often observe a marked difference in grain size in the middle of the dyke, where it has a coarser texture compared to the finer texture or glass at the edges.

The width of the dykes varies from a fraction of a metre to 20 metres: the average being 3.5 m. They are usually several kilometers in length. The longest observed dyke is 21 km long (Suðuroy). On the geological map of the Faeroes, 845 dykes have been indicated.

A dyke which offers greater resistance than the surrounding rock, will stand up above the terrain as a wall (Plate 3.7). The best known example of this is 'Jatnagarðar' on Vágar. If the dyke offers less resistance to erosion than the surrounding rock a gully (Faeroese: gjógv) will be formed (Plate 3.8).

The irregular intrusive bodies have been intruded into the tuff-agglomerate zone and the boundary area between this and the coal-bearing sequence; they occur therefore in the northern part of Suðuroy, Tindhólmur and on Vágar. They are different from the other intrusions with their completely irregular form as a result of intruding a heterogeneous and poorly resistant material such as the tuff-agglomerates. The irregular intrusive bodies can be observed on Suðuroy, on the north and south side of Hvalbiarfjørður, east and south along the coast to the north side of Trongisvágur, where they, on the north side, underlies and invade the tuff-agglomerate zone. They are of considerable dimension and in some places, attain thickness of 100 m.

In the eastern coal area of Suðuroy, between Hvalbiarfjørður and

Plate 3.7. A dyke with eroded side walls, has offered greater resistance to weathering than the surrounding rock.

Plate 3.8. A gully (gjógv) is formed by the erosion of a dyke which was less resistant than the surrounding rock.

Trongisvágur, the coal-bearing sequence have been intruded by irregular intrusive bodies. On Tindhólmur and on Vágar, these intrusives occur under similar condition as on Suðuroy.

Sills follow the originally horizontal bedding plain surfaces so that their extent is parallel with the natural stratification of the surrounding rocks, but it is quite common to see a sill cutting the overlying beds and continue at another level.

The Faeroese sills differ somewhat from other sills in their morphology. They are generally speaking concordant (parallel) with the natural stratification of the surrounding rock, but with increasing transgression towards the periphery. A common characteristic of these sills is that they originally formed saucer-shaped bodies of which the westerly part is now absent. They are thickest in the middle,

Plate 3.9. The Faeroese sills occur in a zone of weakness for the intruding magma between the middle and upper basalt series. The Esturoy sill, as seen above, is 6.5 km long and 2.5 km broad.

often 50 m, and decrease in thickness towards the nearly vertical edges. They often have a clearly developed columnar structure.

The Faeroese sills occur in the boundary area between the middle basalt series and the upper basalt series, which has been a zone of weakness for the intruding magma.

The large Streymoy sill and the large Eysturoy sill are worthy of mention. Apart from these, there are also a number of smaller sills on Streymoy, Eysturoy, Kallsoy, Fugloy and Svínoy. The Streymoy-sill is ca. 9 km long and 1–3 km broad. Eysturoy-sill is 6.5 km long and 2.5 km broad (Plate 3.9). On the west coasts of Streymoy and Eysturoy, these two sills appear as columnar benches with increasing transgression towards the north and the south.

The Faeroese landscape

The preglacial landscape and glaciation. How the preglacial landscape appeared before glaciation we can only imagine, but there can be little doubt that the forces of erosion and weathering have brought their influence as well before as after the glaciation; thus the major terrain forms are considered to be mainly preglacial.

The Faeroes had their own local glaciation, as has already been indicated by James Geikie and Amund Helland in 1880. The glacial striae radiate out in all directions from the centre of the archipelago and foreign erratics are completely absent.

The thickness of the ice is often discussed in earlier literature, it has been suggested that the thickness of the ice had not exceeded 500 m for striae had not been observed above this height. However, it must be taken into consideration that the lack of striae may be due to later erosion of the lava surfaces, which is stronger at higher altitudes than lower altitudes. It is assumed that the ice during the last glaciation extended considerably beyond the present coastline, as remnants of cirques and valley sides commonly occur along the north and west coast. A study of the glacial blocks from the ocean floor around the Faeroes gives the same conclusion.

Major landscape feature. Each of the three basalt series has its own characteristic main landscape form, determined by the rock type, rock structure and the erosive agencies.

In the lower basalt series, the resistant lava benches are interbedded with less resistant tuff and scoria. The erosion has therefore produced a 'trap country', a vegetation covered gently slope changing with steep basalt benches.

The middle basalt series consists of thin lava flows, separated by strongly vesicular zones. The entire series is massive but poorly resistant to erosion and weathering. As these agencies are more active at higher than at lower areas, the removal of eroded material from the higher areas gives rise to characteristic convex slopes in the middle basalt series.

In the upper basalt series basalt- and tuff beds alternate so that the typical 'trap' landscape dominate. As the flows in the upper basalt series however are generally thinner than in the lower basalt series, the 'traps' (steps) are smaller.

Fiords and valleys. As mentioned previously, the Faeroese fiords have a NW-SE trend. This trend corresponds to the system of fissures through which the basalt plateau has been formed. It may be assumed that the system of fissures was caused by tectonic movement prior to volcanism. Preglacial and glacial erosion have since formed NW-SE trending fiords.

Uplift along an axis from the north side of Mykines to the north side of Fugloy had, in the Tertiary, already formed a watershed for the northern islands. North of the axis, the slope is generally ENE and south of it, ESE. The axis follows a gently waving line, crossing the fiords at the shallowest points, and the valleys running in the fiord direction, in the highest point.

During the period of glaciation, this axis has acted as a main ice-shed and it is still acting as a main water-shed.

There are many cirques in the Faeroes; on narrow elongated islands such as Kunoy, Kalsoy and Borðoy, one may observe neighbouring cirques, one facing east and the other west, owning to backward erosion, forming a col.

Morains form a hummocky landscape in the valleys and lower valley slopes. Where the rock are exposed, striae and roches moutonnées give clear evidence of the direction of ice flow. The thickness of the moraine is usually modest, seldom more than 4–5 m.

Mountains constitute an essential part of the Faeroese landscape, particularly in the northern islands which are very mountainous. The parts of the plateau which have been reduced to fragments or 'table mountains', occur at various heights along the longitudinal direction of the islands. When valley sides retreat, one may capture the other and form a crest (Faeroese: kambur); by erosion from more than two sides horn-shaped mountains may be formed (Faeroese: tindur). To the west and north, the mountains may have vertical walls of up to 750 m, but in the interior the slopes become more gentle and are covered with talus.

Lakes of various sizes are common, both in the lower valleys and higher altitudes. Though the largest lakes occur on Vágar, they are most common on Suðuroy, Sandoy, Streymoy and Eysturoy. There are only a few on the northern islands. In previous times there have been more lakes, but several in the lower areas have since been converted into peat bogs. The lakes are partly of glacial origin partly tectonic and a few former bays or creeks.

Examples of lakes of glacial origin can be mentioned: the two largest lakes in the Faeroes, Sørvágsvatn and Fjallavatn, both of which trend NNW-SSE on Vágar. The former, 6 km in length, with an average breadth of 500 m, occurs at 32 a.s.l., and has a depth of up to 50 m. Fjallavatn is approximately 2 km long, 4–500 km wide, lying at 92 a.s.l. and has a depth of 45 m. Other lakes of this age are Leynavatn on Streymoy, Toftavatn and Eiðisvatn on Eysturoy and Kirkjuvatn on Suðuroy.

Lake basins of tectonic origin are formed by erosion in fracture zones; they often appear as a succession of elongated lakes following a straight line or gentle curve.

The lakes Sandsvatn and Gróthúsvatn on Sandoy are two former creeks. Sandsvatn is separated from the sea by about 400 m sandy beach, whilst Gróthúsvatn is separated from the sea by a 50 m barrier of boulders, Sandsvatn lies two metres above sea level, Gróthúsvatn is at four metres above sea level and both are quite shallow.

Peat bogs occur in many places in the Faeroes, not only in the valley floors and lower slopes, but also higher up in the cirques where the floor is peat covered. In some places, particularly on the east side of the islands, the peat bogs extend out under sea level, indicating a recent submergence of the land to the east. The peat horizon is seldom very thick; 1–1.5 m is average and peat horizons deeper than 3 m are seldom.

Shifting sand and dunes. Shifting sands are common near exposed sandy bays but dune formation is uncommon. A dune landscape of modest dimensions, with dunes up to 8 m high, is found on Sandoy on the sandy beach which separates Lake Sandsvatn from the sea.

Gjógv's. A very characteristic feature of the Faeroese landscape is the hundreds of straight, more or less deep, steep-walled gullies which are called, in Faeroese, gjógv. Over 800 have been counted on a 1:20 000 map of the archipelago. They are of tectonic origin and are connected with a lamellar zone system that is younder than the intrusion of dykes.

It is presumed that the dykes were intruded into joints of a lamellar system, whose origin in time was just after the formation of the plateau, while the gjógvs are mainly connected with and eroded in a lamellar zone system that is younger than the dyke intrusion. Gjógvs may vary in size and shape from short shallow gullies to kilometre long, deep gorges. A number of gjógv's may occur along the same lamellar axis.

Coasts. Within fiords and sounds, the coasts are protected and the fiord sides have a gentle slope towards the sea. Where rock is exposed, striae may be visible right down to the waters edge, indicating that erosion has been minimal since glaciation. Outside the fiords however the action of ocean waves forms steep cliffs (Plate 3.10).

Even though the outer coasts are all exposed to similar wave attack, there are marked differences in the morphology of the north and west coasts, compared to the east coast. This is the result of an interaction between several factors. The mentioned difference in rock type and rock structure in the three basalt series will result in different resistance to wave action and produce different morphologies.

Another factor is the dip of the strata with respect to wave action. In places where the strata dips away from the waves (northwest sides), the undermining effect will be much greater than when the strata dip down into the sea (east side).

A further influence is the fact that the east coasts are generally submerged coasts, whereas some sites on the west coasts give clear indications of recent uplift. Finally, fissure and lamellar zone direction is of importance, depending if the zone is parallel or at an angle to the coast.

Along the east coasts, the slope is usually 'stepped' down to the coast. Coastal cliffs higher than 100 m occur only on the northern islands, where the dip is very gentle. Such coasts, with the exception of Suðuroy, are all formed in the upper basalt series.

On the west and north coasts, the form of the high, steep coast dominates, with precipitous, even overhanging promontories up to 750 m (Enniberg). All of these coasts, with the exception of these on Suðuroy, Mykines and the extreme western part of Vágar, are formed in the middle basalt series. This consists of thin lava flows, with vesicular, interbedded zones and tuff which offer much less resistance to wave attack particularly as the strata dip away from wave action. The vesicular interbedded zones provide favourable sites for the undermining wave action.

Plate 3.10. Vertical mountain cliffs (West Streymoy) are formed as a result of the erosion by the ocean waves along the coast.

Plate 3.11. The rapid wave erosion often forms stacks, islets and skerries. This stack is located at Drangarnis, Vágar.

The coast along the steep west sides are strongly affected by rapid wave erosion. Stacks, islets and skerries, are common. Stacks are called 'drangur' or 'stakkur' in Faeroese (Plate 3.11). One of the best known stacks is Trøllkonufingur on Vágar, 313 m.

It is the steep, mountainous coastline on the islands' west and north sides, which are called 'bjørg' (bird mountains). They occur in two forms, 'Meitilberg': high, almost vertical overhanging cliff walls; they are bare or almost free of vegetation with horizontal furrows, ledges and holes formed by the erosion of the porous, interbedded tuff and vesicular zones, 'Vallberg': less steep coastal cliffs with gently sloping, grassy terraces.

The accumulation of blocks, stones and gravels, commonly forms a scree (talus), Faeroese: 'urð', at the foot of slopes; it may be washed away, but if it remains, it will soon be covered with vegetation.

Author's address:
Dr J. Rasmussen
Føroya Natturugripasavn
Torshavn,
The Faeroe Islands

4. Flora and vegetation of the Faeroe Islands

K. Hansen & J. Johansen

Introduction

The flora of the Faeroe Islands, as it occurs, is rather limited. This is, first and foremost, a product of the islands' isolation in the middle of the North Atlantic. The Shetland Islands are the nearest neighbours, ca. 280 km away, and Norway is the nearest mainland at ca. 600 km. It is 320 km to Scotland.

Another reason for the paucity is the modest size of the islands and the homogeneous geological structure which limits the soil types and there with, the number of biotypes.

Vegetation history

The origin of the present flora

The Faeroese flora, at least the majority of it, has invaded the islands since the last ice age. It is likely that some of the highest peaks, for example on Nordoyar, protruded above the ice so that some of the most hardy plants were able to survive, though nothing is known for certain. By far, the most have invaded after the melting of the ice.

The post glacial plant immigration has taken place with ocean currents, by wind and by birds. Aside from this, the first immigrants, both wittingly and unwittingly, brought a number of plants with them to the Faeroes.

The development of the vegetation since the last Ice Age

On the basis of studies carried out by the Geological Survey of Denmark, it is now possible to give a provisional overview of the history of the vegetation (Jóhansen 1971, 1975, 1979).

The earliest flora is not known, as there are no deposits from this period;

however, from about 10 000 b.p., we have a rather clear, though somewhat imperfect, picture of the development.

At the end of the Preboreal, Betula nana grew on the Faeroe Islands. Contemporaneously with and even preceding this, Salix herbacea grew and became more common. At that time, heather, grasses and sedges were also present, but insignificant. Instead, gravel-loving plants grew, such as Sedum, Oxyria digyna, Rumex acetosa, Lycopodium selago and others, which indicate that the soil was rather bare at this time.

Betula nana disappeared at the end of the Preboreal, to be replaced by trees such as Juniperus communis nana and Salix phylicifolia; however, it was grasses and sedges which then became dominant and remained so for the post glacial period, with the exception of the few local areas where local plants were more important.

This development indicates that the climate between the Preboreal and Boreal changed from virtually arctic, with Betula nana, to a more mild, oceanic climate. Juniperus shrub developed on the dry terraces, Salix phylicifolia in the wet depressions and grass moors on the more exposed sites. These condtions prevailed into Atlantic time, although the exact date is not known. Thereafter, the climate became considerably more humid with peat virtually everywhere. Juniperus was depleted at the expense of peat bogs, as Juniperus does not thrive under such conditions. Salix phylicifolia was also somewhat reduced, whilst Calluna vulgaris spread strongly.

In the succeeding Subboreal period, ca. 3000–600 B.C., the climate became drier once again. The peat bogs dried out somewhat, so that Juniperus spread once again. From this period, pieces of Juniperus branches have been found in the Faeroese bogs. In Subatlantic time, from ca. 600 B.C., the climate again became more humid and peat bogs increased. Juniperus was dealt a catastrophic blow, although it did not disappear.

Characteristics of vegetation development on the Faeroes

The development of vegetation on the Faeroe Islands, since the arrival of humans, differs essentially in two respects from most other neighbouring countries.

Firstly, there has not been forest on the Faeroe Islands since the Ice Age. This is due to the oceanic climate, mild winters, extremely variable springs and low summer temperatures.

It appears that the favourable climatic periods have been the Boreal and Subboreal. The intermediate Atlantic period, which in Denmark represents the postglacial climatic optimum, was very humid on the Faeroes.

On Jaeren in southwestern Norway, which also has an oceanic climate, it has been ascertained that the Atlantic period nurtured an unfavourable climate because of excessive precipitation.

The influence of settlement

The Shetland Islands have been inhabited for at least 5000 years. On the Faeroes, no objects older than from Viking times have been found. It is from a written source, however, documented that celibate Irish monks were living on the islands from about 725 until 825 A.D., at which time they were ousted by the Vikings.

In 1971 evidence of a settlement, in Tjørnuvík, between 600–650 A.D. was found. This settlement was probably made by the Irish monks and so it appears that they arrived 100 years earlier than previously thought. Pollen diagrams indicate that land use patterns changed the vegetation.

Cereals have been grown and pastures have been harvested for sheep and cattle. On Mykines in 1979, evidence was found for corn fields from ca. 600 A.D. in which oats (Avena) and later barley (Hordeum) were grown.

The inhabitants brought in new plants to the Faeroe Islands which enhanced the growth possibilities for certain original plants. Amongst the latter are Rumex acetosa and Urtica dioica: both have grown on the Faeroes since Preboreal but increased markedly since human settlement.

The imported plants can mix with the original flora and become naturalized such as Trifolium repens, Vicia cracca and Plantago lanceolata; however, many are not able to compete with the original vegetation. These are not the so-called settlement plants, which can only survive in the immediate vicinity of houses and parks. The homestead plants will be mentioned later.

The flora of the Faeroe Islands

The flora of the Faeroes has been well known for many years. Between 1895–1900, a detailed botanical investigation of the flowering and spore plants was undertaken by E. Warming (1901–1908) and recently (1960–61) there has been undertaken a new, thorough and exact study of the plant distribution on the islands, published by Hansen (1966).

Flowering plants

According to the latest investigations, there are 400 known species of wild or domesticated species of flowering plants; of which 60 species are wild occuring as weeds on cultivated soil or around homesteads. Of the remaining 340 species of flowering plants, the majority are native which are independent of human settlement. A smaller number, around 30 species, were probably imported by humans and, after settlement, spread into the existing plant communities where they now occur, more or less, as natural species.

If one considers only the native species, the Cyperaceae is by far the most species-rich family on the Faeroe Islands (27 species; of which 20 are Carex). This family is certainly the most dominant in the lowlands under 300 m a.s.l. Poaceae with 26 species, is just as dominant in the lowlands as in the higher mountains (over 600 m). The Junacaceae are also well represented on the Faeroes (16 species), particularly on the highlands. It is also certain that the monocotyledones are important components of the Faeroese flora and vegetation.

Other families with many native species belong to the Rosaceae, Caryophyllaceae and Brassicaceae. Finally, it can be mentioned that many species of Saxifraga are common in the highlands.

Most of the plants on the Faeroe Islands are perennial herbs, though a few woody plants do occur; of which, four are bushes which may grow as high as a metre. There are no native free standing trees on the Faeroes as the summer temperature is too low and spring weather too unstable. The four species of bushes are rather uncommon and include the low Juniperus communis nana, Rosa villosa and two low willows: Salix glauca and S. phylicifolia.

As a percentage of all native species, Fanerophyta comprise 0 %, Chamaephyta 15 %, Hemicryptophyta 57 %, Geophyta 13 %, Hydrophyta 10 % and Annuals 5 %. It is particularly characteristic for the highlands of these northerly islands to have relatively many Chamaephyta and only very few Annuals.

Spore plants

Pteridophyta are represented by 27 species, 14 of which are species of Filicales; of these only Bechnum spicant, Athyrium filix-femina, Cystopteris fragilis, Dryopteris filix-mas, and Polypodium vulgare are common on most islands.

There are seven species of Equisetum, of which only Equisetum palustre, E. arvense and E. sylvaticum are common.

There are six species of the Lycopodinae; Lycopodium alpinum and L. selago are common on all islands. Also the little Selaginella selaginoides is found everywhere.

Mosses are a very common part of the Faeroese vegetation. Over 350 species have been found, of which 100 are livervorts. About 20 species are peat mosses; Sphagnum, although they are not nearly as common as in Denmark.

Lichens are an important component of the vegetation, at least on stones, boulders and cliffs. On soil, lichens do not thrive due to the harsh competition with shrubs and mosses, which thrive in the humid climate and the slow growing lichens are reduced or destroyed by the many sheep which wander free in the mountains. The lack of trees excludes, of course, all the species of lichens which only grow on branches and trunks. Of the 250 Faeroese lichens, about 100 are leaf or bush species while the remainder occur as crust on stones.

Approximately 550 species of fungi are known on the Faeroes. In the seas around the archipelago, some 200 species of red, brown and green algae have been distinguished. In Faeroese lakes, some 600 species of algae have been found, the majority of which are microscopic plankton.

Boreal and arctic species

On most of the islands, particularly on Nordoyar, the mountains attain altitudes of 800 m and more. It is therefore obvious that there must be a big difference between the climate in the lowlands and on the mountain peaks, where both the summer, and winter temperatures are lower. Most of the common species in the lowlands can be considered similar to species of northern and western Europe. Of the species on the Faeroes, most are also known in Norway, Sweden and the British Isles; particularly Scotland. These species are generally found everywhere in the lowlands of the Faeroes.

There are also some species that occur mainly, or uniquely, on Suduroy, for example Myosotis secunda, which has a slightly higher annual temperature and somewhat less precipitation than the rest of the islands where the average precipitation is very high; 100–250 cm, three or four times as much as in Denmark.

On the other hand, there are a number of species on the Faeroes whose main distribution pattern falls in arctic areas, Iceland and Greenland, or on the highest mountain peaks in Scandinavia. These arctic species are found primarily only in the highlands, usually above 600 m. Such altitudes are most common in Nordoyar and the northern part of Streymoy and Eysturoy; these arctic species are, therefore, usually limited to these northerly parts of the archipelago. Some examples of these are Gnaphalium supinum, Dryas octopetala, Loiseleuria procumbens, Papaver radicatum, Ranunclulus glacialis and Veronica alpina.

Vertical zonation of the vegetation

Investigations have been made into the vertical zonation of the vegetation. Separate investigations were carried in a lowland zone (under 300 m), an intermediate zone (300–600 m) and a mountain zone (over 600 m — highest peak 882 m). A very clear differentiation was observed between the species distribution in these altitude zones. In the lowland zone, there were 313 species, 180 in the intermediate zone, but only 100 in the mountainous zone. A feature of this clear vegetation zonation is also manifest in a large group of species which occur only in the lowland zone. Important species in this group are: Athyrium filix-femina, Cirsium palustre, Filipendula ulmaria, Juncus conglomeratus and Linum catharticum.

A second group of species occurs almost as commonly in the lowland and intermediary zone but are entirely absent from the mountain zone: Carex echinata, C. panicea, C. pulicaris, Narthecium ossifragum, Prunella vulgaris.

A third group usually occurs in all three altitude zones: Agrostis canina, Armeria maritima, Alchemilla alpina, Luzula spicata, and Thalictrum alpinum.

Finally, there is a group of species which occurs mainly in the mountain zone and seldom in the intermediate zone. Some common examples are the arctic species such as Saxifraga rivularis and Luzula arcuata, which have been previously mentioned.

Coast plant communities

Cliff plants

Most of the Faeroese coastline is comprised of more or less vertical basalt cliffs, commonly several hundred metres in height. The cliffs, at and below water level, are most often thickly covered with species of algae. Zostera marina occurs only in one of the fjords on Suduroy.

The lower parts of the cliff surfaces above water level are usually covered with some crust forming lichens such as Verrucaria, higher up, leaf-like lichens, Xanthoria, and species of the gray, bushy lichen, Ramalina (Plate 4.1). In crevasses and nitches formed by the weathering of the cliffs, coarse textured soil will slowly collect. There is a limited number of species that can tolerate the ocean salt spray and which will take root in such places. Some examples of these are Ligusticum scoticum (Plate 4.2), Matricaria maritima, Cochlearia officinalis, and Plantago maritima.

Beaches and sand dunes

In bays and fjords where waves and currents have less influence, beaches and sand flats of coarse dark sand will be formed by the erosion of the basalt. The plant communities in these areas are quite unique. Nearest the sea, Honckenya peploides and Cakile maritima are the most common: whereas on slightly higher sites, Elymus arenarius, Potentilla anserina and Atriplex glabriuscula all grow.

Only at one place on the Faeroe Islands, at Sandur on Sandoy, are there typical sand dunes behind the beach (Plate 4.3). Here the dominant plant is Ammophila arenaria, accompanied by grasses such as Agropyron junceiforme and Agrostis stolonifera, together with several species from the beach.

Plate 4.1. The lichen Ramalina can tolerate the ocean salt spray and roots in nitches and crevasses.

Plate 4.2. Ligusticum scoticum, a typical coast plant.

The strand meadows

In the innermost part of the fjords and other protected localities, modest strand meadows have been developed about the mouths of small streams which are carrying finer textured material. These strand meadows occur on clay soil which is waterlogged and salty, due to frequent inundation by salt water.

On the lower areas, one finds typical species of Plantago maritima, P. coronopus, Triglochin maritima, Cochlearis officinalis, and Armeria maritima. On slightly higher levels, where salt concentrations are less, one may also find Festuca rubra, Agrostis stolonifera and Leontodon autumnalis.

Plate 4.3. The only sand dunes in the Faeroes. In the foreground Cakile maritima and in the background Ammophila arenaria.

Plant growth in fresh water

Lake plants

The Faeroese lakes are oligotrophic or nutritionally impoverished. The lake strands are almost always gravelly and devoid of plant growth. The gravelly bottom continues a couple of metres into the lake where it falls off steeply; beyond this, the bottom gets soft. The vegetation on the gravel bottom consists mainly of Isoëtes echinospora and I. lacustris, Littorella uniflora, Ranunculus flammula, Subularia aquatica and the characteristic aquatic form of Juncus bulbosus.

Out on the soft bottom, the commonest plants are Sparganium angustifolium, and Myriophyllum alterniflorum, species of Potamogeton, for example: Potamogeton natans, P. gramineus and P. perfoliatus. In addition to these, some species occur from the gravel bottom, particularly species of Isoëtes. Further out, at a couple of metres depth, Charax covers the entire bottom so completely that no other plant can grow there.

There are also other algae in the Faeroese lakes; for example, Cladophora and other green thread-like algae attached to stones in the lake. In addition, there is a large number of microscopic algae. Lake deposits in the Faeroes consist mainly of diatom gyttja, which can be recognized by its light colour and the fact that it does not shrink appreciably when dried. There are also many other microscopic algae, particularly the Desmids — a group of green algae commonly occuring in oligotrophic lakes.

Oligotrophic streams

There are only a few algae and mosses which can grow in river channels where the current is strong. It is only Fontinalis and other mosses, together with some algae, which can survive here. In quieter streams with a more regular flow, the commoner species are Potamogeton polygonifolius and Myriophyllum alterniflorum.

Vegetation in wet sites

Springs are common and can be recognized from a great distance by their bright, light green colour due to mosses: particularly Philonotis fontana, which completely dominates. Higher plants worthy of mention are Saxifraga stellaris, Epilobium lactiflorum, E. alsinifolium and Montia fontana.

Swamp vegetation

There is not very much true swamp vegetation on the Faeroe Islands. It may occur on the banks of a couple of lakes and partly in tarns. A couple of lakes have developed tall vegetation on portions of their banks. These consist of Eleocharis palustris, Equisetum fluviatile, Carex salina and Ranunculus flammula.

Many of the tarns have been completely filled up by Potamogeton polygonifolius, so that there is no open water. Menyanthes trifoliata may protrude its shoots up over the water surface in other tarns: in most of these, submerged species of Sphagnum are important.

Valley marshes

In the bottom of valleys, in depressions and other wet areas, marshes have developed. The basis for this marsh vegetation are the peats, which have played such an important role for the Faeroese population right up to the present time. Turf cutting was very common and has formed artificial depressions in the terrain which has caused even larger areas of this vegetation.

The plants which dominate in this area are Eriophorum amgustifolium, Carex echinata, Carex panicea and Carex nigra. Narthecium ossifragum can locally be of considerable importance as can Molinia coerulea, though this has a somewhat limited distribution.

One notices that Nardus stricta, Juncus squarrous and Scirpus caespitosus occur in the drier parts of the marshes. It is this drier part of the marsh that provides grass pastures for sheep. Juncus squarrosus, with its green rosettes, are particularly attractive to the sheep; in the spring, one notices that practically all of these plants have been eaten down. The other two species are extremely common in drier places and occupy an intermediate position to grass and shrub moors.

Mountain vegetation

Dwarf shrub moors

There are stretches of moors with dwarf shrubs throughout the Faeroe Islands, particularly in the lowlands on the acid and peaty soil which is relatively low in plant nutrients. These dwarf shrub communities occupy an intermediate position between the grass moors of the drier sites and the mountain communities at higher altitudes.

The dwarf bush moors are dominated by low, commonly evergreen bushes; for example, Calluna vulgaris, which is common up to 400 m a.s.l.; Empetrum nigrum and E. hermaphroditum, which may occur even higher up; whilst Erica cinerea is less common and occurs only on moors under 300 m a.s.l.

Other dwarf shrubs worthy of note are Vaccinium myrtillus, which is very common whilst V. uliginosum and V. vitis idaea are less frequent. In these dwarf bush moors, many herbaceous plants are common such as Narthecium ossifragum, Juncus squarrous on the wetter sites and Potentilla erecta, Nardus stricta, Carex pilulifera, Galium saxatile, Hypericum pulchrum and Chamaepericlymenum suecium on drier places. There are also many species of mosses and sometimes species of Cladonia.

The extensive grass moors

The largest part of the surface on the Faeroe mountains consists by far of, more or less, sloping surfaces truncated by vertical cliffs. In the lowlands, 4–6000 m, the slopes will be well drained and relatively dry with a thin humus layer on top of the weathering rock.

The soil here is less acid and more fertile than on the dwarf bush moors, so that the vegetation is dominated by grasses and grass-like plants forming grass moors. These constitute the most important pastures for the thousands of sheep; therefore, they are often strongly grazed. On the other hand, this vegetation is very dense and these considerable areas of grass and other herbs are the essential contributors to the characteristic greenness of the Faeroe mountains when viewed from afar.

The vegetation is composed of different species, of which most occur throughout the islands. The most characteristic grass species are Anthoxanthum odoratum, Deschampsia flexuosa, Agrostis tenuis, Festuca rubra and F. vivipara. Other grass-like herbs such as Luzula multiflora and Carex species are also common.

Finally, the following species are a common component of all grass moors: Leontodon autumnalis, Plantago lanceolata, Prunella vulgaris, Ranunculus repens, R. acris, Cerastium holosteoides and Viola riviniana. In addition, there is a great variety of mosses but lichens are rare as they do not survive sheep grazing very well.

Mountain pastures

As a result of erosion, many flat pasture areas have been formed, from a few hundred metres up to 7–800 m a.s.l. (Plate 4.4). Such plateau-like areas are very exposed to strong winds which, together with high precipitation and humidity,

Plate 4.4. Mountain top plateau, fell field with some lichens, mosses and scattered flowering plants

form a unique soil which supports a sparse vegetation that can tolerate these unfavourable conditions. Many of these also occur in arctic areas.

Most of these areas consist of a mosaic of cliff surfaces with mosses and lichens, with naked gravel, together with a few spread mosses. It is only where the life conditions are favourable, that is in the lee of boulders and on patches of stable soil, where one sees a few species of flowering plants which are usually small and grow in tufts.

The most common plants in these mountain pastures, on most islands, are Silene acaulis, Luzula spicata, Alchemilla alpina, Polygonum viviparum, Juncus trifidus, Koenigia islandica and Deschampsia alpina..

Species of Saxifraga rosacea, Loiseleuria procumbens, Sibbaldia procumbens and the few centemetre high Salix herbacea are similarly typical for the mountain pasture communities in higher places. Finally, Ranunculus glacialis, Papaver radicatum and Gnaphalium supinum occur virtually only on the highest peaks of Nordoyar.

Cliffs and crevasses

Everywhere on the Faeroe Islands, there will be regular vertical and horizontal stairlike slopes on the sides of the basalt mountains which alternate with talus and other slopes. On these vertical cliffs, only plants with strong roots can maintain a foothold in the nitches and loose soil.

The existence of life in such areas depends on the orientation of the cliffs. South slopes will get the most light and heat, but on the contrary, they are more exposed to desiccation, whilst north slopes are dark, cold and wet; these differences will manifest themselves in the vegetation. The vegetation here may appear quite luxuriant; large, strong, richly flowering plants in such places beyond the reach of sheep.

On the Faeroe Islands, there are crevasses with deep, steep sides which may stretch up mountain slopes for many kilometres. These will, of course, present special environments for plant growth.

On the shady, humid slopes of cliffs and crevasses, Oxyria digyna, Saxifraga nivalis, S. rivularis and Sedum rosea are the most common species one sees. Also of significance are Epilobium lactiflorum; the ferns, Dryopteris fili mas and the grasses, Poa alpina and P. glauca.

On the southern slopes, one often finds a luxuriant growth of large plants such as Geranium sylvaticum, Epilobium montanum, Filipendula ulmaria, Luzula sylvatica and many species of Hieracium, Alchemilla and Carex.

Nesting cliffs

The vegetation on the Faeroese nesting cliffs is as unique as the bird life itself. The birds' feces, with their high content of nitrates and phosphates, fertilize the soil and give rise to many tall plants. In addition, these areas are so steep that sheep do not gain access and the vegetation may develop freely. Also sheep will rarely graze such vegetation which is fouled by bird excrements. Such nesting areas are usually divided into three categories: steep cliffs, niches and talus.

The steep cliffs

These are vertical walls where there is not any soil, so that flowering plants can not grow. The only vegetation that can survive are algae, crust-forming lichens and a few mosses. Due to their inaccessibility, the Faeroese nesting cliffs are botanically poorly known, particularly the lower plants.

Niches

Niches are the places where there is some soil and the vegetation is extremely luxuriant. Typical plants are Archangelica officinalis, which are found only on nesting cliffs, and a particularly broad-leaf variety of Festuca rubra, var. fraterculis (in Faeroese, lundasina), which originates from nesting cliffs but has since migrated to other places such as around farmhouses. Where these two species occur, other plants can not compete for light (Plate 4.5).

There are other plants which may occur in the nesting areas: the most common of these are Cochlearia officinalis; Ranunculus acris, which may grow as high as one metre and Matricaria inordòra and Melandrium rubrum; all of which develop strongly.

Talus

Talus consists of the collection of stone and soil at the bottom of steep slopes, in which the many holes and small caverns provide protection for birds.

The plants here are partly the same as in the higher niches; Festuca rubra (fraterculis) and Archangelica officinalis being very important species. Other common plants growing in these areas are Agrostis tenuis, Rumex acetosa, Stellaria media, Cerastium, Sagina procumbens and Urtica dioica, some of which may be found around house sites; the latter likely being a function of higher soil nitrate and phosphate content.

Plate 4.5. Bird cliff on Skúvoy, 300 m high.

Vegetation association with settlement

Enclosed pastures

Approximately 5 % of the total area of the Faeroes (1400 km^2) is cultivated. Previously the main cereals were Hordeum hexastichum and Hordeum distichum. Corn production ceased long ago because the cool and wet summers opposed ripening and drying of the grain. Agricultural land in the Faeroes is now used, almost exclusively, for the production of grass and potatoes. Cultivation procedes by levelling, draining and removing stone from the land. Potatoes are grown for one or two years and the field is then let revert to grass.

Grass seed and other fodder plants are not sewn; the fields are simply allowed to grow by themselves. There are also many useless, although attractive plants in the meadows; therefore, they may instead be called flowering meadows, rather than grass fields. The first invaders after cultivation are Holcus lanatus, Anthoxanthum odoratum, Myosotis discolor, Cardamine hirsuta and others.

At the end of a few years, the fields acquire a certain characteristic; only Anthoxanthum odoratum remains, having displaced the others. Holcus lanatus holds out the longest, but after eight years the following dominate: Anthoxanthum odoratum, Poa sp. and Trifolium repens. Apart from these, there are many useless plants such as species of Ranunculus, Bellis, Prunella vulgaris, Rhinanthus minor, Euphrasia and in the wet areas, Carex sp. and Lychnis flos-cuculi. Faeroese pastures are therefore a colourful site in July when all of these plants flower.

Roof vegetation

Until fairly recently, most houses on the Faeroes have been thatched with grass turf which was laid on a foundation of birch-bark (naever). Even if the turf originally contained many different plants, these gradually disappeared in time so that the thatch vegetation was composed almost entirely of grass (Plate 4.6).

The grasses are best suited for the extreme conditions, particularly drought, which such vegetation is exposed to. The most common are Anthoxanthum odoratum, Poa trivialis and Festuca rubra, var. fraterculis, which are brought from the nesting cliffs. All of these plants develop a luxuriant vegetation so that in some places it may be cut and used as fodder.

Not only grass is growing on the roofs, but Melandrum rubrum and Galium hercynicum are also common. As other roofing is preferred today, these turfing materials have become rare.

Plate 4.6. A farm thatched with turf.

Settlement vegetation

Around house sites, where there is the usual human activity, a number of specific plants thrive. Of particular interest: Rumex longifolius, R. obtusifolius and R. crispus. Poa annua grows around every house and along every road. Apart from these, there is a variety of weeds, such as Capsella bursapastoris, Plantago major, Matricaria matricarioides, Senecio vulgaris and many others.

Garden plants

There is no long tradition of gardening on the Faeroes. When the Danish botanist, Rostrup, visited in 1867, he noticed that there was almost no interest in gardening. Near many houses, there was a little fenced piece of ground where Archangelica officinalis was grown; the fresh tasting stalks were eaten.

In recent years, the interest in gardening has increased so that most new houses have them. With careful attention, many flowers may be grown although a Faeroese garden would have a much poorer appearance than one located further south. Such trees as Sorbus intermedia and S. aucuparia, with Acer pseudoplatanus grow well. Different species of Ribes are also grown.

Establishment of plantations

In 1904, the Danish Moor Association, under the direction of C. E. Flensborg, began tree planting near Torshavn. Broad-leaved trees such as Sorbus intermedia and S. aucuparia, Acer pseudoplatanus, Betula pubescens and Conifers such as Picea abies, P. sitchensis, Abies alba, Pinus mugo, and Larix were planted. Later in the 1920's Pinus contorta, from the Rocky Mountains, was planted amongst these and appears to thrive.

This plantation was successful and the establishment of new plantations have been effected.

Coastal algae

One does not need to follow the coast far to observe a particularly strong and luxuriant algae growth. The situation in the middle of the open ocean, the coastal character and the climate are all factors which are favourable to the growth and thrift of algae.

Large amounts of green, brown and red algae appear. The most common green algae are Enteromorpha intestinalis, Ulva lactuca and Cladophora rupestris.

Brown algae probably catch the eye the most as they cover extensive areas and several species attain considerable dimensions; of these, Fucus spiralis, and Fucus vesiculosus are worthy of mention. In addition, appreciable amounts of Himanthalia elongata, Laminaria digitata, Laminaria hyperborea, and Alaria esculenta may be found. Ascophyllum nodosum is quite commonly distributed.

The red algae also have a wide distribution, the most common being Lithothamnion polymorphum, Gigartina stellata, Corallina officinalis, and Chondrus crispus. Some of the red algae are particularly attractive, especially those which grow in deep waters.

Author's addresses:
Mr K. Hansen
Føroya Natturugripasavn
Torshavn,
The Faeroe Islands

Mr J. Johansen
Føroya Natturugripasavn
Torshavn
The Faeroe Islands

5. Animal life on the Faeroe Islands

D. Bloch

Land animals

In considering the land fauna on the Faeroe Islands, there are two essential factors which have led to the selection of animals. The first is the isolation with a water barrier of 675 km to the nearest mainland. The other factor is the comparative geomorphological youth. The older the islands are, the greater the likelihood that more animals have crossed the water barrier and established themselves. Of the groups that can neither swim nor fly, the amphibians and reptiles are entirely absent and the few wild mammals have all arrived with humans.

Mammals

The three known land mammals are the mouse (*Mus musculus*), the brown rat (*Rattus norvegicus*) and the snow hare (*Lepus timidus*). The snow hare has been brought in on purpose and the other two are involuntary immigrants with humans.

Mice and rats

Mice and rats supplement each other in distribution in that they both occur together only in the Torshavn area. Otherwise rats occur on all the large islands except Sandoy where mice dominate. In addition, there are mice on some of the smaller outermost islands. The mice can be divided into four different forms; on Fugloy, Mykines, Nólsoy and Torshavn, which are each inhabited by its own form. The forms on Mykines and Nólsoy are especially markedly divergent. They are large and have strong back legs and tails, which is probably an adaption to their life on the bird nesting cliffs. Mykines-mice remind one of the mice on St. Kilda, the outermost island in the Hebrides. This may be due either to a similar milieu or to common origin.

Monographiae Biologicae, Vol. 46 ed. by G. K. Rutherford

Rats and the large gulls are considered to be the islands' worst vermin. When one wanders in the snow during the winter, one sees from the rat tracks just how common they are. Winter, when food is lacking, would be the most convenient time to regulate their population. When the birds return in the spring and begin to lay eggs, there is soon an excess of food. Rats are dispersed as far as the coastal nesting cliffs in summer and are a considerable mortality factor amongst the breeding Puffin (*Fratercula arctica*) and Manx shearwater (*Puffinus puffinus*).

The Faeroese hare

This hare was introduced in 1855 for hunting, and two pairs were set loose in the Torshavn area. They were imported from southern Norway where two common species of hare come together: *L. t. timidus* and *L. t. canescens* (Plate 5.1).

L. t. timidus has a completely white winter coat and gives birth to two litters a year whilst *L. t. canescens* is bluegrey in winter and bears three litters annually. In addition, it has shorter ears, mates earlier and sheds its coat later. The imported hares were originally white in the winter, but as early as 1882 — after only 28 years — only 25 % were white, whilst the rest had a blue-grey winter coat.

The last white hare was shot in the winter 1916–17. Now the Faeroese hares have three litters annually.

It has been shown using experimental animals in cages, that the blue-grey winter colour is due to a recessive factor which one supposes was present in one of the animals that were imported in 1855. The interplay between the modest original population has caused the greater fertility of *L. t. canescens*, and the poorer camouflage colour of *L. t. timidus* to dominate so that the dominant genes have been suppressed in the course of 60 years.

The hare spread rapidly and by 1900 could already be found on some other islands. There is open season in November and December, with the yield usually quite large.

Bird life

The landscape of the Faeroes is not particularly varied and the natural food production is sparse; the bird life is correspondingly limited. On the other hand, these birds which nourish themselves from the oceans bounty, occur in large numbers. The limiting factor for sea birds would be the lack of nesting places, whilst for the land birds it would be the lack of food.

There are other things which can influence bird population; rats limit population numbers particularly for Puffins (*Fratercula arctica*) and Manx shearwater (*Puffinus puffinus*) (Plate 5.2). An increase in the numbers of the large gulls, such as the Herring gull (*Larus argentatus*) the Lesser Black-backed gull (*L. fuscus*) and the Greater Black-backed gull (*L. marinus*), constitute a major imposition on such land birds as the Eurasian Golden plover (*Pluvialis apricaria*) and the Whimbrel (*Numenius phaeopus*). The number of gulls, Ravens (*Corvus corax*) and Carrion crow (*C. corone*) has increased greatly in the last ten years, associated with the increase of refuse sites.

The hatching sites for ducks and the Red-necked Phalarope (*Phalaropus lobatus*) are becoming dangerously limited as these are being drained for hydro power. Whilst nesting cliffs for sea birds have long been protected, there is as of yet not a single local inland area which has been protected for biological reasons. On the other hand, the Faeroese hunting regulations are quite superior as they protect in principle all animals with the few exceptions mentioned in the regulations.

Resident birds

The Faeroes is the Nordic country which has the relatively highest number of sedentary birds. This is due to several variables. The mild winter climate with an average temperature of 4 °C for the coldest month in Torshavn provides a possibility for birds which are migratory in e.g. Norway and Sweden to spend the winter here. Another factor necessary for a migratory bird is to be able to return to its nesting place. A little island group such as the Faeroes can be difficult to find again, and the risk of faulty navigation is high. Evidence shows that some species will survive better as a nonmigratory bird.

Amongst the birds which are migratory in Norway and/or Sweden, but non-migratory on the Faeroes, the Mallard (*Anas platyrhynchos*), the Black-headed gull (*Larus ridibundus*), the Common gull (*L. canus*), the great Black-backed gull (*L. marinus*), and the Carrion crow (*Corvus corone*) are all worthy of mention.

During the course of time, some birds have become stationary here and developed into special Faeroese subspecies. Some of these include the Starling (*Sturnus vulgaris faeroensis* Feilden), Wren (*Troglodytes troglodytes borealis* Fischer), the Water Pipet (*Anthus spinoletta kleinschmidti* Hartert), the Black Guillemot (*Cepphus grylle faeroensis* Brehm) and the Eider (*Somateria mollisima faeroensis* Brehm).

There are also Faeroese subspecies among the migratory birds which are isolated in the winter quarter, such as the Guillemot (*Uria aalge spiloptera* Salomonsen). In addition there are a number of subspecies which are common to Iceland and the Faeroes, such as the Merlin (*Falco columbarius subaesalon* Brehm), Water rail (*Rallus aquaticus hibernans* Salomonsen), the Snipe (*Gallinago gallinago faeroensis* Brehm), the Oystercatcher (*Haematopus ostralegus malacophaga* Salomonsen), Redshank (*Tringa totanus robusta* Schiøler) and the Redwing (*Turdus iliacus coburne* Shape).

Plate 5.1. The Faeroese hare originated from Norway in 1855, spreading rapidly. They are now found on several islands. (Photo: Kalmar Lindenskov.)

Plate 5.2 The population of the Puffin is often limited due to the presence of rats. (Photo: Arne Nörrevang.)

Nesting birds

It is natural to divide the bird of the Faeroese into the land birds and sea birds associated with the coastal nesting cliffs. Approximately 53 species nest regularly in the Faeroes.

Land birds

One can divide these birds into categories of the lakes, the remote areas, the birds of the near fields and birds in areas of settlement. *Lake birds.* There are many lakes in the Faeroes ranging from small tarns to the major Sørvagsvatn of 3.4 km^2. They are all low in nutrients, clear, with sparse bottom vegetation. In addition, most of them occur in the remote open fields where sheep limit the bank vegetation. Lake birds are correspondingly few in numbers.

The most common species is the Mallard (*Anas platyrhynchos*) and the Red-breasted Merganser (*Mergus serrator*). Both of these ducks nest evenly about the islands. The Mallard (*Anas platyrhynchos*) is the commonest of the two, but on the other hand, the Red-breasted Merganser (*Mergus serrator*) is seen at sea all the year round, and in the winter some of them are migrants from the north. Among the nesting birds, which are few in number, is the Red-throated Diver (*Gavia stellata*). There are only some 5 to 10 pairs and they occur only on the calmest lakes. In addition, there are a few pairs of Horned Grebe (*Podiceps auritus*). Red-necked Phalarope (*Phalaropus lobatus*) can be seen in several places. This strange little bird has a pelagic life in the winter and may be difficult to discover in its nesting place. It may be seen in small ponds and streams connected to larger lakes, and is one of the birds whose nesting habitat is endangered.

In the summer, one may catch a glimpse of the Tufted Duck (*Aythya fuligula*), the Scaup (*Aythya marila*) and the Common Scoter (*Melanitta nigra*) on the lakes and perhaps even a few pairs hatched.

Birds of the open fields

The hill country where the 70- 80 000 sheep spend all the year, can be roughly divided into three different types of environment.

The dwarf bush moor seldom goes above 300 m and occurs on east facing valley slopes. These acquire more precipitation than the west flanks which are in a rain shadow area for the west winds, the dominant wind direction. As well, there is more sunshine on the west flanks.

Grass heaths occur in the drier areas and at higher altitudes, both of these types give way to virtually naked rock. In the highest areas one may see a few

Purple Sandpiper (*Calidris maritima*), which is usually rather trustful, and the large Ringed Plover (*Charadrius hiaticula*); even less common, one occasionally can find Snow Bunting (*Plectrophenax nivalis*) one or two pairs of nesting. In the stony higher sites, the Eider (*Somateria mollissima*) may be found, but there are also nests on smaller islands and on grassy screes. The Water Pipit (*Anthus spinoletta*) is quite common at the highest altitudes and is replaced lower down by the Meadow Pipit (*Anthus pratensis*). However both of these species will overlap in the nesting areas, and particularly the Water Pipit (*Anthus spinoletta*) may be seen everywhere.

Highest on the mountain sides and in crevasses by the coast, one may find the nesting places for the Rock Doves (*Columba livia*) which have increased in latter years. In rocky walls on the hillsides ('Hamrar'), the nests of Raven (*Corvus corax*), Carrion Crow (*Corvus corone*) and the Fulmar (*Fulmarus glacialis*) can be found. Lower down in the grass and dwarf shrub moors, there are three characteristic birds. First and foremost, there is the Faeroe Island national bird, the Oystercatcher (*Haematopus ostralegus*) and also the Whimbrel (*Numenius phaeopus*) and the melodious Golden Plover (*Pluvialis apricaria*). In summer, one can not move in this area without having at least one of these three flying around as you pass through their territory.

Here one finds the islands' only bird of prey, the Merlin (*Falco columbarius*), which appears to be increasing in numbers in latter years. It nests both in the mountains and in the 'Hamrar'. Finally there are several birds which nest in colonies. Of particular interest is the Arctic Tern (*Sterna paradisea*) which one may find in colonies both inland and on the coast. The colony usually spends only a few years in the same place. In these areas, one can also find colonies of the Arctic Skua (*Stercorarius parasiticus*) and the Great Skua (*S. skua*). Both of these birds are hunted all year, and particularly the Great Skua (*S. skua*) has decreased markedly since 1972 when it was no longer protected.

In the lowest part of this area, on fences and stones, the Wheatear (*Oenanthe oenanthe*) makes his home. It is a sure sign of spring when one sees it hopping around the stones sometime in April.

In winter, this area may be inhabited by a number of migrating and stationary Snow Bunting (*Plectrophenax nivalis*) Fieldfare (*Turdus pilaris*) and Redwing (*T. iliacus*).

Birds of the home fields and settlement areas

There are no particular birds associated with the home fields. One can happen upon a number of Snipes (*Gallinago gallinago*) in drainage ditches and in wet areas. Otherwise it is mainly flocks of foraging birds that one sees and these may be residents, sites of migratory birds or winter guests. There are always flocks of Starlings (*Sturnus vulgaris*) which in winter are supplemented by Turnstone

(*Arenaria interpress*). In early spring when the Oystercatcher (*Haemotopus ostralegus*) returns about the 12th of March they sit in flocks and forage before they disperse in pairs to their various territories. Later in the autumn, when these fields have been cut for hay, great flocks of the Oystercatcher are to be seen, now accompanied by their flying young together with some Black-headed Gulls (*Larus ridibundus*) and Common Gulls (*L. canus*). Around houses with fenced gardens, one sees Starlings (*Sturnus vulgaris*), House Sparrows (*Passer domesticus*) and Wrens (*Troglodytes troglodytes*) throughout the year. In summer, one may notice a White Wagtail (*Motacilla alba*) or a Swallow (*Hirundo rustico*) which are becoming more and more common, and throughout the year the Robin (*Erithacus rubecula*).

In larger bushes and trees around houses one may hear two kinds of thrushes, the Blackbird (*Turdus merula*) and the Redwing (*T. iliacus*), both of which have increased in numbers in recent years.

Since 1920 a number of plantations have been established on the islands, the oldest being in Torshavn. On this plantation in particular, one finds many Collared turtle Doves (*Streptopelia decaocto*). This bird has spread rapidly since it arrived in the Faeroes in 1968. In harbours one can almost always see the Eider (*Somateria molissima*), and perhaps the Shag (*Phalacrocorax aristotelis*) and in the winter time a flock of Turnstone (*Arenaria interpress*).

Birds of the coastal nesting cliffs

Just as the inland areas are scarce in birds, the coastal areas are correspondingly rich. The birds associated with the coastal cliffs are nourished by the sea and this enormous food resource is a result of the archipelago's position in the ocean currents. Around the Faeroes, the warm, low nutrient Gulf Stream meets the cold, higher nutrient water from the Arctic. These waters mix here and set the scene for a very high production of algae from April to September. These in their turn provide nourishment for vast amounts of zooplankton which nourish both fish, birds and humans. In the winter, food production is minimal and the nesting cliffs inhabitants have long been dispersed. Even so, a few birds remain here over winter so that one may encounter the coming nesting season's birds such as Fulmar (*Fulmarus glacialis*) from November, and in a similar way, the last of this year's nesting birds Storm Petrel (*Hydrobates pelagicus*) may remain until New Year (Plates 5.3).

At sea, one may see at this time Guillemot (*Uria aalge*), Little Auk (*Plotus alle*) and less commonly the Great Northern Diver (*Gavia immer*).

The birds have thus made the most of the island's potential. The nesting cliffs begin to fill up in April and May, and already about the first of August, they begin to empty again. Highest in the coastal cliffs, but also in the 'Hamrar', one may note the Fulmar (*Fulmarus glacialis*) and following the cliffs down towards the ocean one finds on the first ledge Guillemot (*Uria aalge*) and Kittiwake (*Rissa tridactyla*). In screes at the foot of the nesting cliffs, Shag (*Phalacrocorax aristotelis*), Black Guillemot (*Cepphus grylle*) and Razorbill (*Alca torda*) may all nest (Plate 5.4).

Colonies of Puffin (*Fratercula arctica*) may always be found away from the true nesting cliffs on sloping lands towards the ocean. Out on Mykines-Holmur, colonies of Gannet (*Morus bassanus*), and the Leach's Petrel (*Oceanodroma leucorrhoe*) occur.

Fowling

Birds have been hunted for centuries partly at sea and partly at the nesting cliffs. The importance of fowling is evident in the hunting regulations where 24 of the 26 major paragraphs are concerned with fowling. Today there are only a few species of birds that are of importance for fowling. On the island of Mykines, which is now almost uninhabited, a few Gannets (*Morus bassanus*) are taken (Plate 5.5). Guillemot (*Uria aalge*) and Razorbill (*Alca torda*) are much less common in the last ten years and got protected for five years ahead in May 1980. A few young Fulmar and Fulmar eggs are taken, and some nesting Manx Shearwater (*Puffinus puffinus*).

Numerically the hunting of Puffin (*Fratercula arctica*) is only of importance today. This bird can only be caught by 'Fleyging' which is the use of a four meter long racquet-like net; thousands are caught each year for home consumption.

Terrestrial invertebrates

The invertebrate fauna are generally poorly known. There are a number of Earthworm (*Lumbricidae*) and even high up, and amongst the Beetles (*Coleoptera*), the large *Carabus problematicus* is most common. On the coast one finds the large Thysanyra (*Petrobius brevistylis*). Altogether there are some 43 species of Spiders (*Araneae*) and 300 species of Flies (*Diptera*).

Grasshoppers (*Saltatoria*) and Ladybirds (*Coccinellidae*) were registered a few years ago, while complete groups like the Moths (*Lepidoptera*), Dragonflies (*Odonata*), Mayflies (*Ephemeroptera*) and Stone Flies (*Plecoptera*) are lacking. One certainly sees flying Butterflies once in a while. Small Tortoiseshell (*Aglais urticae*), Peacock (*Inachis io*), Red Admiral (*Vanessa atalanta*) and Tidselfugl? (*Vanessa cardui*). The Flies (*Diptera*) dominate pollenation, and Crane-Flies (*Tipupidae*) can be found in large numbers, and their larvae do a great deal of damage in pastures.

Plate 5.3. The Fulmar often remains on the nesting cliffs during the winter, even though food is mimimal. (Photo: Arne Nörrevang.)

Plate 5.4. One often finds Guillemots on the ledge of the nesting cliffs. (Photo: Arne Nörrevang.)

Plate 5.5. Gannets are fowled on the Faeroes, often on the island of Mykines. (Photo: Arne Nörrevang.)

Fresh water fauna

The many different lakes varying from drained tarns to large deep and cold water lakes represent the possibility for a rich fauna, if poor in species. A certain nutrient supply to the lakes is brought about by the deposition of guano. Such lakes have numerous Water Fleas (*Daphniae*) and some twenty species of Caddis Flies (*Tricoptera*) and 80 species of 'Dansemyg' (no English name) (*Chironomidae*) have been registered. Gnats (*Culicadae*) are quite absent, but there are several species of Horse Flies (*Simuliidae*).

Fish

There are only five different fresh water fish which all are euryhaline. Stenohaline fresh water fish have not had any possibility to reach the Faeroes on their own.

Eels. (*Anguilla anguilla*) are found in most waterways. The small Elvers occur in summer and they swim into the brackish water at river mouths or further up into fresh water. They grow slowly in the cold water as it takes 10–20 years until they become silver eals when they leave the Faeroes late in the autumn to migrate to the southwest.

Salmon. (*Salmo salar*) and Trout (*Salmo trutta*) are found as well. It is not known if the salmon was original or not, but each year, smallfry are set out so that it is now a native stock. In the same way, trout are set out for sports fishing. Both the Sea Trout (*S.t. trutta*), which inhabits the salt and fresh water, and the fresh water Brown Trout (*S.t. fario*) occur.

In a single deep lake (69 m) a relic from the ice age has been found, called a Charr (*Salvenius alpinus*). This has been transferred to other lakes where it has grown quite big. Finally one finds the very euryhaline Three-spined Stickleback (*Gasterosteus aculeatus*). The little fish may be encountered both in the ocean, along the coast, in estuaries, in rivers and high up in fresh water lakes.

Animal life in the ocean

The bountiful harvest of the ocean is a lifeline for the Faeroes where fishing is by far most important occupation. The position of the Faeroes on the submerged ridge between Scotland and Iceland means that the area of the land plinth is much bigger (ca. 21 000 km^2) than the land area (1400 km^2). Going from the coast towards the ocean, the bottom falls off quickly to about 15 m and thereafter it slopes gently to a depth of 200 m.

In order to describe the marine life it would be natural to begin at the coast and work out.

The coastal zone

The animal life along the coast is very dense partly because of the strong algae growth and partly due to stones, rock and cliffs presenting favourable footholds for the animals. One sees many snails; Common Limpet (*Patella vulgata*), Common Dog Whelk (*Nucella lapillus*), The Rough Periwinkle (*Littorina saxatilis*) and Flat Periwinkle (*L. obtusata*). The latter is found only in protected fiords and sounds.

In addition, there are many Common Mussels (*Mytilus edulis*) and the Horse Mussel (*Modiolus modiolus*). The newly secreted mussel larvae may be precipitated onto sea weed and late in the year, around 700 animals per square metre of seawood may be found in the protected fiords. Of these, 96% may be less than 4 mm long. Amongst the low water crustaceans, one finds many barnacles (*Balanidae*), particularly the common barnacle (*Balanus balanoides*) and among the many free living shell fish one sees, the Isopods (*Idotea granulosa, I. baltica* and *I. pelagia*) together with the shore Crab (*Carcinus maenas*).

The seaweed belt is also rich with fish fry eggs particularly of species such as saithe (*Pollachius virens*), Cod (*Gadus morhua*), Dab (*Limanda limanda*) and Plaice (*Pleuronectes platessa*). Occasionally one finds fish whose habitat is the seaweed belt and ebb tide pools such as the Three-spined Stickleback (*Gasterosteus aculeatus*). The Bull Rout (*Acanthocottus scorpius*), which is common in the northerly parts, whereas the long Sea Scorpion (*A. bubalis*) is common in the south. In this area one might also find examples of Montagu's Sea Snail (*Liparis montagni*) and Sea Snail (*L. liparis*).

Bottom fauna

In sounds and fiords which are protected from the Atlantic swell, one usually finds a fine clay bottom with a corresponding bottom fauna. In the lowest sites, from 10–15 m depth, the chalky Macoma (*Macoma calcarea*) is characteristic. One also finds here, the clams *Syndosmya nitida, Thyasira flexuosa* and *Mya truncata*, as well as the sedentary Annelid worms *Nepthys ciliata* and *Goniada maculata*.

Out in the deeper water from about 25 to 80–90 m, the area is characterized by the little clam *Thyasira flexuosa*; *Macoma calcarea* may be spotted sporadically. This bottom area is also inhabited by the clams (*Syndosmya nitida, Nucula tenuis and N. nucleus, Nuculana pernula*). Amongst the snails, one finds the

Pelican's foot (*Aphorrhais pes-pelicani*) and amongst the sedentary bush worms, *Stenapsis fossor* is very common and *Praxilella praetermissa*, *Terebellides strömi*, *Goniada maculata* and *Nepthys cialata* also occur. In this area, one also finds the Sea Cucumber *Cucumaria frondosa* and the Sea Urchin (*Strongylocentrotus droebachiensis*).

The lobster (*Nephrops norvegicus*) also lives out here in this clay bottom community. Approximately 40 tons of these are fished every year. Amongst the fish which are particularly associated with the soft bottom are Dab (*Limanda limanda*), and Long Rough Dab (*Hippoglossoides platessoides*); and the more important Plaice (*Pleuronectes platessa*), Lemon Sole (*Microstomus Kitt*), Witch (*Glyptocephalus cynoglossus*) and Halibut (*Hippoglossus hippoglossus*). One finds as well the small examples of Haddock (*Melanogrammus aeglefinus*) and Whiting (*Merlangus merlangus*).

Coarse bottom fauna

On exposed cliffs, out to 70–80 m in depth, one finds a community of sedentary invertebrates characterized by Dead Man's Fingers (*Alcyonium digitatum*). The Bivalves (*Saxicava arctica*, *Heteranomia squamula*) and *Chiton*, together with the Brush Worm (*Filigrana*) are very common.

Further out in the coast, at a depth of 80–110 m, one finds a fauna dominated by the Horse Mussel (Modiolus modiolus), often with the Barnacle (*Balanus hammeri*). Other animals which are plentiful here are the Starfish, Brittle Star, Crabs (*Galathea nexa*, *Spirontocaris pusiola*, *Eupagurus bernhardus*) and Polychacks (*Lipidonotus Squamatus* and *Harmotroe imbricata*).

Most of the animals listed above are important fish food, especially the Cod (*Gadus morhua*). This animal community is also found out on the lower parts of the Faeroese banks. From this community, 500 tons of Queen Scallops (*Chlamys opercularis*) were fished in 1978.

Between 110 and 300 m depth, there is a sand bottom community characterized by the Elliptical Trough Shell (*Spisula elliptica*) but the clams (*Oval Venus*, *Striped Venus*, *Tellina pygmea*, *Gari tellinella and Lima subauriculata*) are also common. In this community, one may also find the Brittle Stars (*Ophiura albida* and *O. sarsi*) and the Sea Urchins (*Brisaster fragilis*, *Spatangus purpureus*) as well as *Echinocyamus pusillus*.

This community also occurs out on the Faeroe Banks. The two communities described here as Modiola and Spisula, constitute the bottom fauna in most of the areas where professional fishing takes place.

Fish

By far the most important fish is Cod (*Gadus morhua*) of which 31 000 tons were fished in 1978.

Two independent Cod stocks occur, the one around the Faeroes and the other out on the Faeroese Banks. The Bank Cod are characterized by growing much quicker than the Faeroese Cod, as in four years it weighs 4.1 kilos and measures 80 cm, whilst the Faeroese Cod, at four years weighs 2.6 kilos and is 61 cm long. Another kind of Codfish with an independent Faeroese stock is the Haddock (*Melanogrammus aeglefinus*). The haddock fishery has increased greatly in the last 15 years and that almost 20 000 tons were taken in 1978.

Another important fish, as yet mostly used for fish meal and oil production, is the Blue Whiting (*Micromesistius poutassou*), of which 200 000 tons were fished in 1978.

Some other important commercial fish which are taken from around the Faeroes include the Saithe (*Pollachius virens*), 28 000 tons landed in 1978; Ling (*Molva molva*), 4500 tons in 1978; Tusk (*Brosme brosme*), 3500 tons landed in 1978.

In addition, some Halibut (*Hippoglossus hippoglossus*) and Lemon Sole (*Microstomus kitt*), comprising 1–2000 tons, are fished annually.

For some time it has been considered that Cod and Haddock are being overfished so that a quota arrangement has been introduced in order to increase stocks.

A few whales are shot each year, mostly Common Rorqual (*Balaenoptera physalus*), but it is the hunting of the Pilot Whale (*Globicephala melaena*), which is most important as some 1–2000 are taken each year.

Author's address:
Ms. D. Bloch
Føroya Natturugripasavn
Torshavn,
The Faeroe Islands

6. Species assemblages and coexistence of Faeroe Island Ground Beetles (Coleoptera: Carabidae)

S.-A. Bengtson

Introduction

The Faeroes, situated in the North Atlantic between the Shetland Islands and Iceland (approx. 62°N, 7°W), consist of 18 basaltic islands ranging in size from 1 to 374 km^2 (total area: 1400 km^2) and the highest peaks of the different islands reach 370–880 m NN. The Faeroes are oceanic in so far as they have not been connected with continental land masses in post-glacial time. The climate is characterized by frequent and strong winds and high and frequent precipitation (often >2000 mm and rainfall on 260 days of the year). The difference between the mean temperatures of the coldest and warmest months is only about 7 °C (4.1 °C in February and 11.1 °C in August). The vegetation cover is dominated by various types of grass heaths and grass meadows, but on some south and west facing slopes, areas of dwarf shrub heaths occur and on higher grounds (>250 m NN) patchily vegetated fell-fields are found. Thus, at first sight, the climate and vegetation may suggest environmental uniformity and stability. However, as will be argued below, studies of the terrestrial invertebrates (here Carabidae) suggest rather variable environmental conditions.

Qualitatively, the Ground Beetle fauna of the Faeroes is well studied; mainly through extensive field-work by Danish zoologists in the mid-1920's [West 1930] and by a Swedish-Norwegian team in 1978–1979 (present study). West [1930] lists 26 species belonging to the family Carabidae (for names and nomenclature see Table 6.1) of which 18 were also found by us; the remaining 8 species seem to be either rare (or accidental) or mostly confined to seashore habitats, which were less thoroughly investigated. At least 15 species of Carabidae appear to be relatively widely distributed and common in the Faeroes (present study) and the faunal composition is entirely NW European [West 1930].

This paper gives a descriptive account of the Faeroese Ground Beetles with references to geographic and habitat distribution within the island group. However, the major part of the paper attempts to relate field-data to modern ecological concepts such as species-abundance relationships, species-packing, resource partitioning, limiting similarity, and principles of coexistence. These

Monographiae Biologicae, Vol. 46 ed. by G. K. Rutherford

Table 6.1. Species of Ground Beetles [Coleoptera: Carabidae] recorded on 17 of the islands in the Faeroes. Records denoted (+) refer to collections made in 1978–1979 and (*) to earlier published records [West 1930]. Columns to the right give occurrence for each species: number of islands (including*), number of localities where pitfall-trapping was carried out (n = 78), and number of additional localities surveyed in 1978–79 (n = 33), where the figures in brackets refer to data found in West [1930]. The nomenclature follows Lindroth [1974].

Species	Islands																	Number of		
	Súduroy	Skúvoy	Stóra Dímun	Sandoy	Mykines	Vágar	Koltur	Hestur	Nólsoy	Streymoy	Eysturoy	Kalsoy	Kunoy	Bordoy	Vidoy	Svinoy	Fugloy	Islands	Localities with pitfalls	Additional localities
Carabus problematicus Herbst 1786	*				+	+			*	+	+			+	+	+		9	12	6
Nebria gyllenhali Schönherr 1806	+		+	+	+	+	+	+	+	+	+	+	+	+	+	+	+	16	63	21
Nebria salina Fairmaire & Laboulbene 1856	+	+		+	+	+	+	+	+	+	+	+	+	+	+	+	+	16	70	21
Notiophilus aquaticus (Linnaeus 1758)										+	+							2	3	(2)
Notiophilus biguttatus (Fabricius 1779)	+	+	+	+	+	+	+		+	+	+	+		+	+	+	+	15	39	13
Loricera pilicornis (Fabricius 1775)	+			+	+	+	+			+	+	+	+	+	+	+	+	13	37	4
Patrobus septentrionis (Dejean 1828)	+		+	+	+	+	+	+	+	+	+	+	+	+	+	+	+	16	74	18
Patrobus assimilis Chaudoir 1844	+			+		+			+	+	+		+	+				8	36	1
Patrobus atrorufus Ström 1768)	*			+		+			+	+	+			+				7	8	1
Trechus obtusus Erichson 1837	+	+	+	+	+	+	+	+	+	+	+	+	+	+	+	+	+	17	70	10
Trechus fulvus Dejean 1831	+					*			*	*								4	1	(6)
Bembidion bipunctatum (von Linné 1761)	*			*		*				*	*							5		(4)
Bembidion geniculatum Heer 1838[a]	*			*						*								3		(5)
Bembidion bruxellense Wesmael 1835[b]	*									*								2		(2)
Pterostichus adstrictus Eschscholtz 1823	*			+		+			*	+	+	+		+	+			9	17	1
Pterostichus nigrita (Paykull 1790)	+			+		+	+			+	+			+	+			8	19	3
Pterostichus diligens (Sturm 1824)	*								+	+	+							4	5	
Pterostichus strenuus (Panzer 1796)										*								1		(1)

Table 6.1 (continued).

Species	Islands																	Number of		
	Suduroy	Skúvoy	Stóra Dímun	Sandoy	Mykines	Vágar	Koltur	Hestur	Nólsoy	Streymoy	Eysturoy	Kalsoy	Kunoy	Bordoy	Vidoy	Svinoy	Fugloy	Islands	Localities with pitfalls	Additional localities
Calathus fuscipes (Goeze 1777)	+		+	+	+	+			+	+	+		+	+	+	+	+	13	42	6
Calathus melanocephalus (Linnaeus 1758)	+	+	+	+	+	+	+	+	+	+	+	+	*	*	+	+	+	17	31	5
Olisthopus rotundatus (Paykull 1790)	*																	1		(1)
Amara aulica (Panzer 1796)	+			+	+	+			+	+	+							7	22	1
Harpalus quadripunctatus Dejean 1829	*			*														2		(2)
Dicheirotrichus gustavi Crotch 1871					*				*									2		(2)
Trichocellus cognatus (Gyllenhal 1871[c]	*			*	*	*			*	*								6		(7)
Trichocellus placidus (Gyllenhal 1827)	+									*								2	1	(2)
Number of species 1978–79:	13	4	6	13	10	14	8	5	11	16	16	8	7	12	10	9	8	(18 species)		
Total number (incl.*):	23	4	6	17	12	17	8	5	16	23	17	8	8	13	10	9	8	(26 species)		

[a] *B. redtenbacheri* K. Daniel 1902.
[b] *B. rupestre* auct., nec (Linnaeus 1767).
[c] *D. pubescens* (Paykull 1790) /praeocc./.

concepts are interrelated and often difficult to disentangle, but are important to ecological theory and help to elucidate mechanisms which allow species to coexist. Current literature generally assumes that populations usually exist in equilibria with their resources and that competition is of paramount importance in determining patterns of coexistence [see Pianka 1976]. This may well be so, although it is difficult to obtain conclusive field evidence for competition, and recently the basic assumptions of resource-defined equilibria have been questioned [Wiens 1977]. So, which patterns of coexistence can be detected among Faeroese Carabidae and what do they suggest?

Material and methods

The field-work was carried out in July-August 1978 and in the first half of August 1979. Altogether 17 of the 18 islands (not the smallest of them; Lítla Dímun) were visited and 78 localities (Fig. 6.1) were surveyed by using pitfall traps (with 4 % formalin), which were run for an average of about three weeks at each site. In addition, most of the localities were investigated by taking sieve samples which were subsequently extracted in Tullgren funnels, and by intensive hand-collecting. Another 33 localities were surveyed only by sieving and hand-collecting. It is well-known that, since the catches depend on the activity of the animals, pitfall traps may often give a biased picture of species composition and relative numbers of individuals [for a review regarding Carabidae, see Thiele 1977].

As for the qualitative aspects of our study, it turns out that sieving and hand-collecting added only 9–10 % of the species to the 78 species list obtained by pitfall sampling alone. Considering the quantitative aspects, we did not detect any consistent of significant differences between pitfall catches and sieve samples with regard to relative abundance of the species. Presumably pitfall-trapping is an adequate method for comparing different sites in the Faeroes, where distances between localities are usually small and structural differences between habitats are minor (e.g. no vertical zonation of the vegetation).

Since the field-work was not primarily designed to study Ground Beetle ecology (but human influence on several taxa of terrestrial invertebrates), the choice of sampling localities does not exactly mirror the proportional occurrence of habitats in the Faeroes. All islands are strongly dominated by grass heaths and, around the many settlements, various types of grass meadows. The field-work was focussed on a comparison between habitats within and around villages and farmsteads, i.e. the «infields» or «homefields» (böur in Faeroese) which are usually surrounded by fences (often stonewalls) to exclude sheep, and habitats outside the fences («outfields» or hagi in Faeroese) where sheep roam freely. Each locality was assigned to one of the following nine habitat types:

A. Mountain sites (250–650 m NN), or fell-fields, with sparse and patchy vegetation including *Silene acaulis* (L.) Jacq., *Luzula spicata* (L.) DC.,

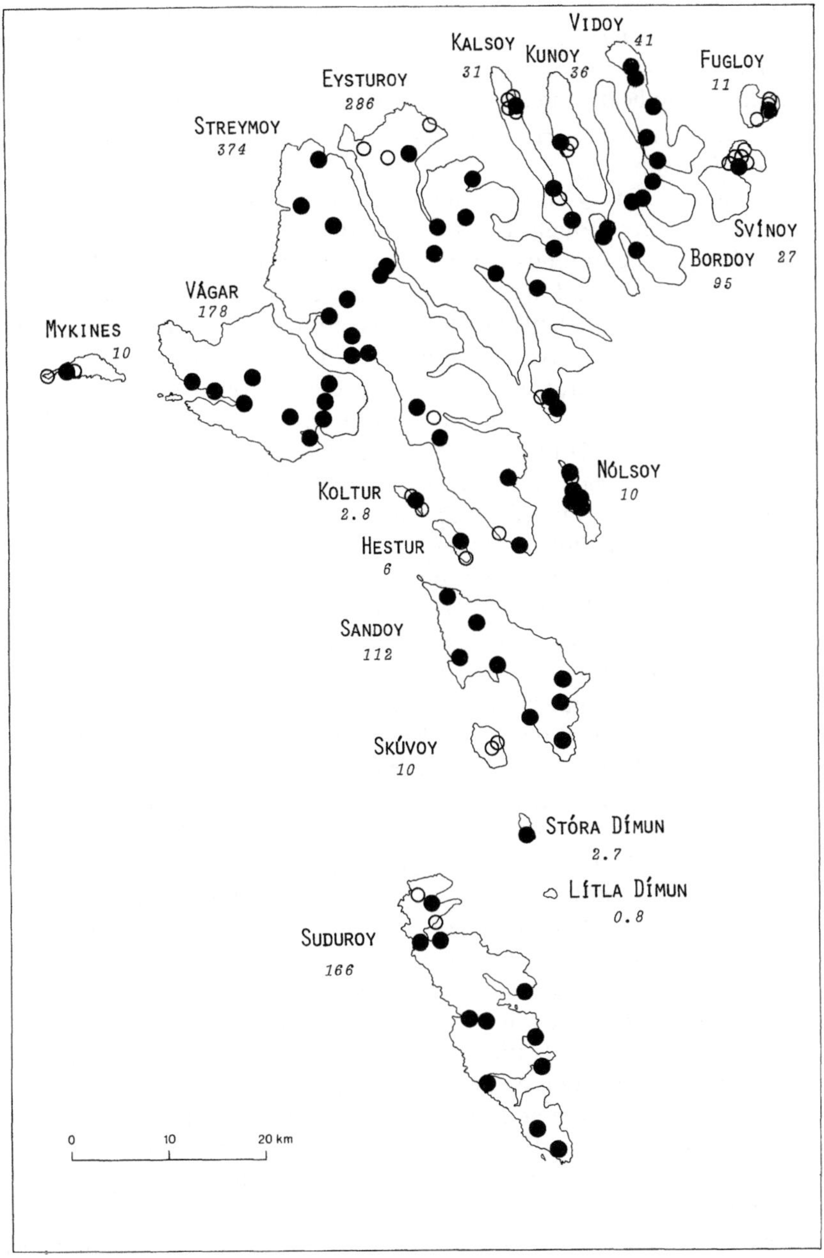

Fig. 6.1. Map of the Faeroes giving the names of the islands, size in km^2, and position of localities surveyed in 1978–1979. Black circles (n = 78) refer to localities with, and open circles (n = 33) to localities without pitfalls.

Alchemilla alpina L., *Nardus stricta* L., *Deschampsia alpina* (L.) Roem. & Schult., *Polygonum viviparum* L., *Juncus trifidus* L., *Salix herbacea* L., and mosses of *Rhacomitrium* spp.

B. Dwarf shrub heaths on south and west facing slopes with a dominance of *Calluna vulgaris* (L.) Hull, but also *Erica cinerea* L., *Empetrum* spp, and *Vaccinium* spp.

C. Plantations, mostly small (a few hectares or less), consisting mainly of coniferous trees though with some deciduous trees such as *Sorubs*, *Acer*, and *Betula*. The ground-layer is usually absent or poorly developed. The plantations are found in a few places only and within or in the vicinity of villages. They were planted early in this century and in the 1950's. Natural woods have never grown in the Faeroes in post-glacial time.

D. Cliffs, shelves (hamrar in Faeroes), and crevices (gjaír in Faroese) which are inaccessible to the sheep and therefore have a lush vegetation including *Luzula sylvatica* (Huds.) Gaudin, *Angelica archangelica* L., *Geranium sylvaticum* L., *Sedum rosea* (L.) Scop., and *Epilobium* spp.

E. Lowland bogs dominated by *Sphagnum*-species, various *Carex* spp, and *Eriophorum angustifolium* Honck.

F. Grass heaths (<250 m NN) of different types, dominated by species such as *Festuca rubra* L., *F. vivipara* (L.) Sm., *Anthoxanthum odoratum* L., *Agrostis tenuis* Sibth., *Deschampsia flexuosa* (L.) Trin., *Ranunculus* spp, *Prunella vulgaris* L., *Leontodon autumnalis* L., and many mosses. This is the most common habitat in the islands and sheep-grazing is extensive.

G. Infield grass meadows on the outskirts of settlements with a rich and varied vegetation including *A. odoratum*, *Holcus lanatus* L., *Trifolium repens* L., *Bellis perennis* L., *Rhinanthus minor* L., *Euphrasia officinalis* L., *Myosotis discolor* Pers., *Lychnis flos-cuculi* L. and many others. Sheep are excluded from infield meadows used for haymaking.

H. Infield localities inside or near villages (gardens, graveyards etc) with an often very luxuriant vegetation of grasses and herbs, including a number of ruderal species such as large *Rumex* spp.

I. Sand dunes, with *Ammophila arenaria* (L.), Link, *Elymus arenarius* L., and *Cakile maritima* Scop, which are found only in one place on Sandoy.

The number of localities (with pitfall traps) from each habitat is given in Table 6.2. For further details on the localities and collecting efforts see Bengtson & Hauge [1979]. As to the nomenclature and for more information about the flora and vegetation of the Faeroes it is referred to Hansen [1966].

The pitfall catches were converted to number of individuals caught per trap and unit time ('activity abundance') and the results are considered to be indicative of habitat utilization. This information was used to calculate niche breadths (β) according to Levins' [1968] formula:

$$\log \beta = -\sum p_i \log p_i$$

Table 6.2. Number of species and species diversity (the Shannon-Wiener function) of Ground Beetles [Coleoptera: Carabidae] in different habitats (cf. text) in the Faeroes. Only pitfall samples are included (n=78).

Habitat	No. of localities	Number of species Min.	Max.	Mean (±SD)	Mean species diversity (±SD)
(A) Mountain sites	2	4	7	5.5 (±2.1)	1.273 (±0.234)
(B) Dwarf shrub heaths	7	5	11	8.0 (±2.1)	1.462 (±0.453)
(C) Plantations	2	5	8	6.5 (±2.1)	1.212 (±0.132)
(D) Cliffs, shelves & crevices	1			7	1.204
(E) Lowland bogs	3	4	7	5.7 (±1.5)	0.816 (±0.260)
(F) Grass heaths	19	4	9	6.3 (±1.4)	1.184 (±0.433)
(G) Infields: outskirts	24	5	11	7.9 (±1.7)	1.299 (±0.313)
(H) Infields: settlements	19	3	10	6.8 (±1.9)	1.280 (±0.250)
(I) Sand dunes	1			2	0.488

where p_i is the proportion of individuals of a given species using habitat i. The same data were used to calculate the overlap in habitat use between species a and b according to Pianka's [1974] formula:

$$\alpha_{ab} = \frac{\sum_{i}^{n} p_{ia}\, p_{ib}}{\sqrt{\sum_{i}^{n} p_{ia}^{2} \sum_{i}^{n} p_{ib}^{2}}}$$

Thus, habitat is one resource dimension assumed to be partitioned among the species. Food is assumed to be another dimension for which, however, no adequate data are available. All species of Carabidae [except *Amara aulica* which to a large extent feeds on plant material] are presumably carnivorous though, like most Ground Beetles, they undoubtedly also eat plant material [see Thiele 1977]. Differences in body-size are often assumed to indicate dissimilarity in food ecology (types of food and feeding strategies) and resource utilization along the food dimension. The assumption that morphometric data reflect differences in food ecology has not always been verified [see e.g. Hespenheide 1973, Wilson 1975, Turner & Polis 1979]. Nevertheless, I have examined differences in total body-length (which is strongly correlated with body-breadth and mandible-size), measured from random samples of individuals. The 15 common species fall into the following five arbitrary size-categories (body-length):

Very small [3.9–5.3 mm]: *Notiophilus biguttatus* [5.0 mm]
Trechus obtusus [3.9 mm]
Pterostichus diligens [5.3 mm]

Small [6.2–7.9 mm]:	*Loricera pilicornis* [7.0 mm]
	Patrobus assimilis [7.9 mm]
	Calathus melanocephalus [6.2 mm]
Medium-sized [9.5–11.3 mm]:	*Nebria gyllenhali* [10.2 mm]
	N. salina [10.8 mm]
	Patrobus septentrionis [10.3 mm]
	P. atrorufus [9.5 mm]
	Pterostichus adstrictus [11.3 mm]
	P. nigrita [9.6 mm]
Large [12.9–13.3 mm]:	*Calathus fuscipes* [13.3 mm]
	Amara aulica [12.9 mm]
Very large [21.2 mm]:	*Carabus problematicus*

Clearly, it would have been of considerable interest to know the types, and relative abundance, of food available in different localities and habitats. Such information is extremely difficult to obtain; I have used the amount of animal matter (g dw/unit time) collected in the pitfall traps as a very coarse «food-index» to rank the abundance of prey available to the Ground Beetles in different localities. To avoid possible bias owing to differences between years I have used only the 70 pitfall series collected in 1978.

Species distribution and constancy

Table 6.1 lists all 26 species of Carabidae known to have been found in the Faeroes. As can be seen, ten species have been recorded on more than half the number (i.e. 9) of the islands. One group consists of six species, viz. *N. gyllenhali*, *N. salina*, *T. obtusus*, *P. septentrionis*, *C. melanocephalus*, and *N. biguttatus*, which have been found on all (17) or nearly all (≥ 15) islands. A second group, consisting of *L. pilicornis* and *C. fuscipes*, are almost as widely distributed, and a third group of six species, viz. *C. problematicus*, *P. assimilis*, *P. atrorufus*, *P. adstrictus*, *P. nigrita*, and *A. aulica*, occur on 7–9 islands. The remaining 12 species appear to be more locally distributed, rare, or accidental.

Species with a wide distribution also show high constancies (i.e. occur in a high proportion of the localities surveyed) and, for instance, *P. septentrionis*, *T. obtuses*, and *N. salina* were found in 70–74 of the 78 pitfall-trap localities and in many of the additional 33 localities (Table 6.1). Some other species, e.g. *P. assimilis*, *P. nigrita*, and *A. aulica*, with a less wide distribution, also show high constancies. There are few tendencies towards geographic patterns in the interisland distributions: *A. aulica* and *P. atrorufus* appear to be absent from the northeastern islands (i.e. Kalsoy, Kunoy, Vidoy, Svínoy, and Fugloy).

Species-area relationships

The 17 islands range in size from 2.5 km^2 (Stóra Dímun and Koltur) to 374 km^2 (Streymoy) and the number of Ground Beetles vary from 4 to 23 on the different islands (Table 6.1). The low numbers found on Skúvoy (4) and Hestur (5) most certainly reflect incomplete and less thorough surveys. The number of species is, as expected, positively correlated with island size (all 17 islands: $r=0.76$ and 15 islands: $r=0.78$; in both cases $p<0.001$). The species-area relationships can be described by the common equation: $S=CA^z$, where S is the number of species, C is a constant, A is area, and z the slope of the log-log regression. For all 17 islands the equation reads $S=4.54\ A^{0.25}$ and for 15 islands (excluding Skúvoy and Hestur) $S=5.67\ A^{0.21}$. The z-values are within the range usually found in similar studies [Connor & McCoy 1979].

Species number and diversity

The number of species found at a single locality varied between 2 and 11 and the range of variation within each of the habitats was almost as large (Table 6.2). The highest mean number of species was found in infield localities on the outskirts of settlements (7.9) and on dwarf shrub heaths (8.0), but all other habitats contained on average almost as many species (5.5 to about 7), and only the sand dune locality (two species) differed markedly from the rest (Table 6.2). The number of species per locality is significantly correlated ($r=0.42$; $p<0.001$) with increased prey-abundance, although within-habitat variation is great (Fig. 6.2).

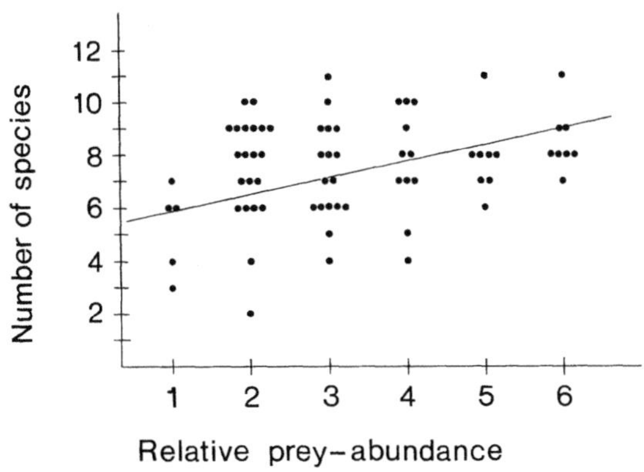

Fig. 6.2. The positive relationship (see text) between number of species of Ground Beetles [Coleoptera: Carabidae] and relative prey-abundance (i.e. a food-index calculated from pitfall catches of potential prey) in the Faeroes. Only pitfall localities surveyed in 1978 are included ($n=70$).

Species diversity (using the Shannon-Wiener function) is generally low, varies from place to place, and shows no consistant pattern with regard to habitat; except that it is low in sand dunes and bogs and relatively high on dwarf shrub heaths (Table 6.2). However, there is a statistically significant positive correlation between species diversity and food-index ($r=0.37$; $p<0.001$). This also is true when size-classes (5 classes, see §2) are plotted against abundance of prey, i.e. the food-index ($r=0.44$; $p<0.01$).

Species assemblages and relative abundance

Species composition and relative abundance (dominance) in the nine habitats are presented in Table 6.3. Several species were found in many different habitats but the dominant species were always found among the following: *P. septen-*

Table 6.3. Relative abundance (% dominance) of Ground Beetles [Coleoptera: Carabidae] in different habitats in the Faeroes. For each habitat pitfall samples from several localities were pooled (cf. Table 6.2). + denotes < 0.1%. Capitals refer to: (A) mountain sites, (B) dwarf shrub heaths, (C) plantations, (D) cliffs, shelves, and crevices, (E) lowland bogs, (F) grass heaths, (G) infields: outskirts, (H) infields: settlements, and (I) sand dunes. For further information see text.

Species	Habitat								
	A	B	C	D	E	F	G	H	I
P. septentrionis	29.8	49.9	6.8	50.0	78.8	54.6	35.9	16.8	–
T. obtusus	34.6	9.6	14.1	35.0	3.9	16.0	10.5	11.3	–
N. gyllenhali	11.5	0.7	10.7	1.8	1.5	4.7	10.6	45.4	80.9
C. fuscipes	–	10.8	–	–	–	1.5	10.8	13.2	19.2
N. salina	1.9	5.8	42.4	3.6	5.4	5.1	4.8	4.9	–
C. melanocephalus	13.9	1.0	–	1.8	–	0.6	8.2	2.8	–
N. biguttatus	7.7	2.4	0.6	6.8	0.5	4.0	0.2	0.7	–
P. atrorufus	–	–	23.2	–	–	–	6.3	1.6	–
P. assimilis	0.5	9.6	–	0.9	3.9	4.5	5.4	1.2	–
P. adstrictus	–	2.9	–	–	1.0	4.5	0.4	+	–
L. pilicornis	–	–	1.7	–	1.5	0.9	2.2	1.6	–
P. nigrita	–	2.9	–	–	3.5	1.3	0.5	–	–
C. problematicus	–	3.9	–	–	–	0.4	0.6	0.1	–
A. aulica	–	–	0.6	–	–	–	3.7	0.5	–
P. diligens	–	0.5	–	–	–	1.5	+	+	–
N. aquaticus	–	–	–	–	–	0.3	–	–	–
T. fulvus	–	–	–	–	–	–	+	–	–
T. placidus	–	–	–	–	–	0.1	–	–	–
Number of individuals	208	415	177	220	203	1522	3119	3488	47
Number of species	7	12	8	7	9	15	16	14	2

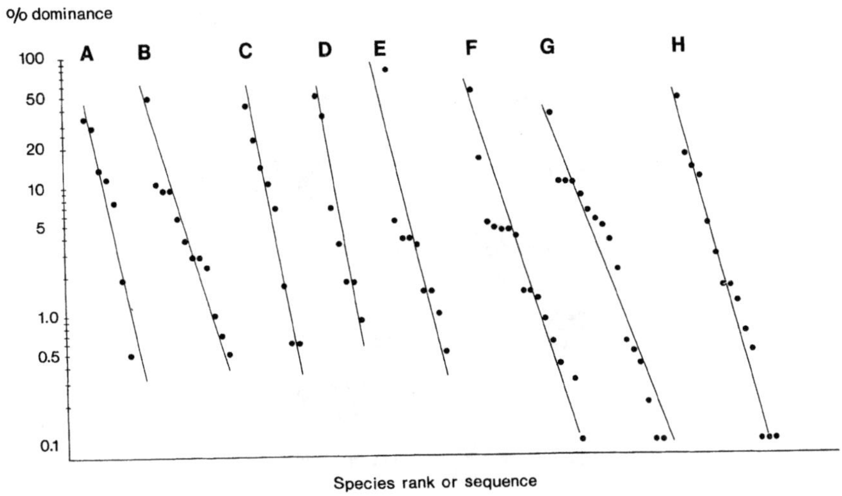

Fig. 6.3. Ranked species-abundance curves for assemblages of Ground Beetles [Coleoptera: Catabidae] in 8 different habitats (A–H; cf. Table 3) in the Faeroes. The regression lines are fitted by eye.

trionis, *T. obtusus*, *N. gyllenhali*, and *N. salina*. In mountain sites *P. septentrionis* and *T. obtusus* dominated, followed by *C. melanocephalus* and *N. gyllenhali*. On dwarf shrub heaths *P. septentrionis* was the dominant species with *C. fuscipes*, *T. obtusus*, and *P. assimilis* as sub-dominants. In plantations *N. salina* dominated, followed by *P. atrorufus* and *T. obtusus*. On cliffs, shelves and in the crevices *P. septentrionis* and *T. obtusus* were the dominants, and the former was by far the most abundant species on lowland bogs. On grass heaths *P. septentrionis* alone dominated, followed by *T. obtusus*. Infield localities included most of the species and both in and around, and on the outskirts, of settlements *P. septentrionis* was abundant, though in «settlement-localities» *N. gyllenhali* dominated. In both infield-categories *T. obtusus* and *C. fuscipes* were sub-dominants. On sand dunes only *N. gyllenhali* (the commonest) and *C. fuscipes* occurred.

When the dominance values in Table 6.3 are used to construct species-abundance curves it appears that, for each habitat, the data best fit a logarithmic series (Fig. 6.3).

Habitat niche breadths

Habitat utilization spectra of the 15 most common species are shown in Fig. 6.4. All species occur in several different habitats: five species in eight habitats, ten species in ≥ 5 habitats, and none in < 3 habitats. Calculating the habitat niche

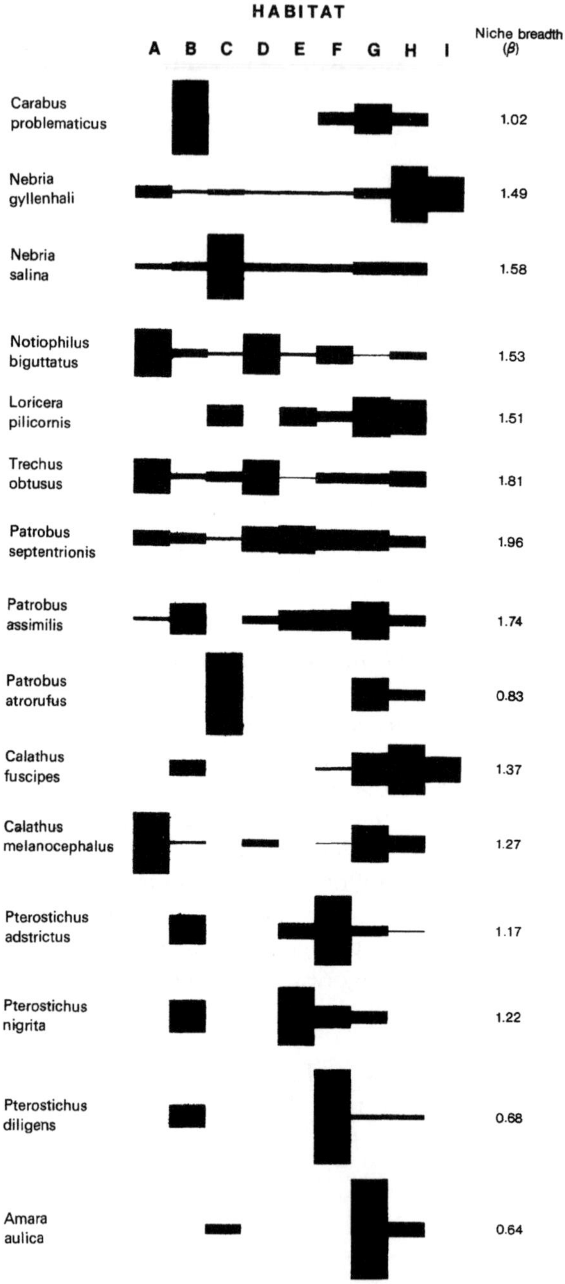

Fig. 6.4. Habitat utilization and habitat niche breadths (β) of Ground Beetles [Coleoptera: Carabidae] in the Faeroes. The habitat spectra describe the proportion of individuals of given species found in different habitats based on number of individuals caught in pitfall traps per trap and unit time. The habitat abbreviations are; (A) mountain sites, (B) dwarf shrub heaths, (C) plantations, (D) cliffs, shelves, and crevices, (E) lowland bogs, (F) grass heaths, (G) infields: outskirts, (H) infields: settlements, and (I) sand dunes (see also text).

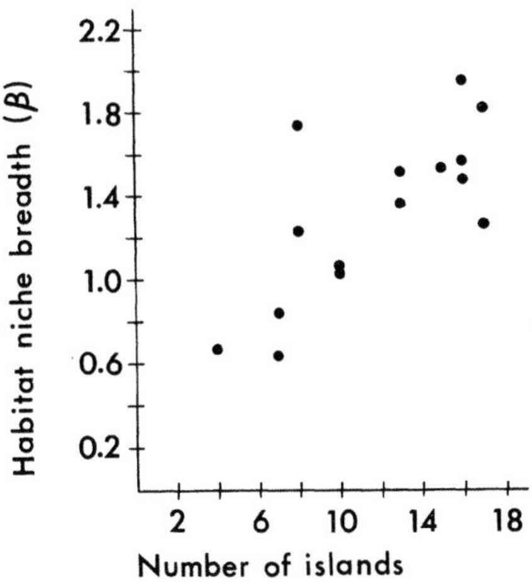

Fig. 6.5. Relationships between habitat niche breadth (cf. Fig. 6.4) and number of islands on which a species occurs (cf. Table 6.1) for Ground Beetles [Coleoptera: Carabidae] in the Faeroes.

breadths (β) shows that *P. septentrionis*, *T. obtusus*, and *P. assimilis* have the broadest niches (1.74–1.96), followed by a group consisting of *N. biguttatus*, *L. pilicornis*, *N. salina*, and *N. gyllenhali* (1.49–1.58), and thereafter a group including *C. fuscipes*, *C. melanocephalus*, *P. nigrita*, and *P. adstrictus* (1.17–1.37). Relatively narrow habitat niche breadths are exhibited by *C. problematicus* (1.02), *P. atrorufus* (0.83), and especially by *P. diligens* (0.68), and *A. aulica* (0.64). Habitat niche breadths are positively correlated with the number of islands on which the species are found ($r = 0.75$; $p < 0.001$); though it is noteworthy that *P. assimilis* occurs on fewer islands than its habitat niche breadths would predict (Fig. 6.5).

Niche overlap

Interspecific overlap (α) in resource utilization is calculated for two dimensions, viz. habitat and food ecology (inferred from body-size ratios), for 15 species of Faeroes Carabidae (Table 6.4). Already a close overall inspection of Fig. 6.4 suggests that the species are well segregated by habitat; this is also supported by the calculated α-values which have an overall mean of 0.37 (Table 6.5); this must be considered a low degree of overlap. However, there is considerable variation within the matrix ranging from very low overlaps of <0.05 [7 cases: *N. gyllenhali* vs *P. nigrita*, *N. biguttatus* vs *P. atrorufus*, *N. biguttatus* vs *A. aulica*, *P. atrorufus* vs *P. nigrita*, *P. atrorufus* vs *P. diligens*, *C. melanocephalus* vs *P. diligens*, and *A. aulica* vs *P. diligens*] to high overlaps of >0.89 [4 cases: *N. gyllenhali* vs *C. fuscipes*, *N. salina* vs *P. atrorufus*, *N. biguttatus* vs *T. obtusus*,

Table 6.4. Matrix showing habitat utilization overlap (α) and body-length ratio (largest: smallest species) of 15 common species of Ground Beetles [Coleoptera: Carabidae] in the Faeroes.

	C. problematicus	N. gyllenhali	N. salina	N. biguttatus	L. pilicornis	T. obtusus	P. septentrionis	P. assimilis	P. atrorufus	C. fuscipes	C. melanocephalus	P. adstrictus	P. nigrita	P. diligens	P. aulica
C. problematicus		0.13	0.17	0.16	0.30	0.21	0.41	0.78	0.14	0.45	0.23	0.52	0.52	0.34	0.35
N. gyllenhali	2.08		0.31	0.23	0.62	0.40	0.31	0.28	0.25	0.89	0.39	0.07	0.05	0.07	0.28
N. salina	1.96	1.06		0.21	0.60	0.39	0.34	0.28	0.95	0.25	0.18	0.18	0.20	0.13	0.28
N. biguttatus	4.24	2.04	2.16		0.14	0.95	0.69	0.31	0.05	0.12	0.70	0.29	0.17	0.29	0.04
L. pilicornis	3.03	1.46	1.54	1.40		0.40	0.61	0.69	0.63	0.76	0.46	0.29	0.41	0.20	0.76
T. obtusus	5.44	2.62	2.77	1.28	1.80		0.76	0.43	0.27	0.33	0.76	0.25	0.22	0.23	0.26
P. septentrionis	2.07	1.01	1.05	2.06	1.47	2.64		0.79	0.21	0.40	0.55	0.57	0.73	0.41	0.41
P. assimilis	2.68	1.29	1.37	1.58	1.13	2.03	1.30		0.26	0.59	0.46	0.67	0.72	0.48	0.68
P. atrorufus	2.23	1.07	1.14	1.90	1.36	2.44	1.08	1.20		0.27	0.21	0.04	0.06	0.01	0.48
C. fuscipes	1.59	1.30	1.23	2.66	1.90	3.41	1.29	1.68	1.40		0.42	0.18	0.19	0.13	0.59
C. melanocephalus	3.42	1.65	1.74	1.24	1.13	1.59	1.66	1.27	1.53	2.15		0.09	0.10	0.05	0.53
P. adstrictus	1.88	1.11	1.05	2.26	1.61	2.90	1.10	1.43	1.19	1.18	1.82		0.63	0.96	0.11
P. nigrita	2.21	1.06	1.13	1.92	1.37	2.46	1.07	1.22	1.01	1.39	1.55	1.18		0.39	0.16
P. diligens	4.00	1.93	2.04	1.06	1.32	1.36	1.94	1.49	1.79	2.51	1.17	2.13	1.81		0.03
A. aulica	1.64	1.27	1.19	2.58	1.84	3.31	1.25	1.63	1.36	1.03	2.08	1.14	1.34	2.43	

Body-length ratio (below diagonal); Habitat overlap (α) (above diagonal)

Table 6.5. Mean body-length ratios and habitat overlap (α) of different categories of Faeroese Ground Beetles [Coleoptera: Carabidae].

		Number of species	Mean size-ratio (±SD)	Mean habitat overlap (±SD)
Size-class	4–5.5 mm	3	1.23 (±0.16)	0.49 (±0.40)
	6–8 mm	3	1.18 (±0.08)	0.54 (±0.13)
	9.5–11.5 mm	6	1.09 (±0.05)	0.33 (±0.28)
	13–14 mm	2	1.03	0.59
Patrobus spp		3	1.19 (±0.11)	0.42 (±0.32)
Pterostichus spp		3	1.71 (±0.48)	0.66 (±0.29)
All species		15		0.37 (±0.24)

and *P. adstrictus* vs *P. diligens*]. When mean overlaps within size-classes are calculated it seems as if the medium-sized species are less densely packed than the other classes (Table 6.5); though the differences are not statistically significant. There are three species each of *Patrobus* and *Pterostichus* but the overlaps in habitat utilization within these genera (0.42 and 0.66) are not significantly higher than the overall average (Table 6.5); thus, overlap is moderate.

Body-length ratios range from 5.44 [*C. problematicus* vs *T. obtusus*] to a number of cases close to 1.00 (Table 6.4). In many instances the ratios are well below 1.26, which was Hutchinson's (1959) observed coexistence-value.[1]

Table 6.6. A simplified matrix (cf. Table 6.4) comparing 15 Faeroese Ground Beetles [Coleoptera: Carabidae] with regard to body-length ratios (largest: smallest species) and overlap in habitat utilization. The figures give the number of species-pairs (total 105) for each of the nine combinations. The delimitation of low, moderate, and high is arbitrary; except for the size-ratio 1.26 which follows Hutchinson [1959].

Body-length ratio	Habitat overlap (α) low <0.30	moderate 0.30–0.60	high >0.60	
Low < 1.26	14	7	6	n = 27
Moderate 1.26–1.80	15	11	10	n = 36
High > 1.80	23	13	6	n = 42
	n = 52	n = 31	n = 22	n = 105

General principles of coexistence predict that species which show a high degree of overlap with regard to one resource dimension must be well separated along other (complementary) dimension to enable to coexist. Table 6.6 summarizes the information contained in Table 6.4 by arbitrarily grouping the α-values and body-length ratios into categories (low, medium, and high), allowing each of the 105 «species-pairs» to be assigned to one of nine possible combinations of two-dimensional niche overlap. It turns out that 83 of the species-pairs are segregated by a low or moderate habitat overlap whereas 78 are segregated by body-length ratios > 1.26. The number of «unsolved cases» (i.e. a high habitat overlap combined with a low body-length ratio) include only six pairs: *N. biguttatus*, vs *C. melanocephalus*, *P. atrorufus* vs *N. salina*, *P. assimilis* vs *L. pilicornis*, *P. assimilis* vs *P. nigrita*, *P. septentrionis*, vs *P. nigrita*, and *P. adstrictus* vs *P. nigrita*. Although these species-pairs appear to be too similar in their ecologies to coexist, they were found together in a number of localities that almost exactly equalled the numbers expected from their constancy values and randomness. Also, attempts to use the number of individuals caught per unit time (as an index of population density) in order to detect possible inter-specific interferences failed to indicate effects on numbers or distribution of any species.

[1] However, the sizes of the Faeroese Ground Beetles cannot be distinguished from a model of independent, random, uniform arrangement between largest and smallest by the Barton-David test for all ratios ($G_{1.14} = g_1/g_{14} = 0.021$, $0.10 < p_r < 0.90$), the Darling test for the maximum ratio ($g_{14} = 0.275$, $p_r(g_{14} > 0.275) = 0.211$), and the Irwin test for the minimum ratio ($g_1 = 0.006$, $p_r(g_1 < 0.006) = 0.067$) (Prof. Daniel Simberloff pers. comm.). — [Added in proof].

Discussion

It has become axiomatic that islands support fewer species than comparable continental areas of land (for a theoretical background see MacArthur & Wilson (1967), which is also true for Faeroese Carabidae. The historical background of the Ground Beetle fauna of the Faeroes is inadequately known.

There is subfossil evidence of *Calathus micropterus* (Duftschmid 1812) (now apparently extinct in the Faeroes), *P. nigrita*, and *C. problematicus*, dating back to subboreal or subatlantic time (Jessen & Rasmussen 1922). Presumably, much more species remain to be discovered in the peat of the Faeroes. The islands have no endemic Carabidae; only *C. problematicus feroensis* Lapouge 1910 holds subspecific rank [see Lindroth 1968]. The lack of endemics and subspecies may, however, be of little significance when discussing the age of the Ground Beetle fauna because the family seems to have evolved relatively slowly during Pleistocene (Thiele 1977: 303). West [1930] suggests that a part of the present fauna may have reached the islands via a pre-glacial continental land-bridge and survived the Würm glaciation (s) on nunataks which may have occurred in the Faeroes; at least there is no contrary geological evidence. This view was supported by Lindroth (1932), a strong adherent of the importance of glacial refugia in helping to explain present-day biotas in various parts of the Northern Hemisphere. Although it is clear that many species of Carabidae must have colonized the islands before the arrival of man (the earliest evidence is from 600–650 A.D. according to Jóhansen (1971; pers. comm.)), it is equally clear that man and his activities must have played an important role in moulding the present Faeroese fauna; as for NW European biotas in general [see Brinck 1966]. Indeed, already long ago Lindroth [1931: 515–516] pointed to the marked differences in the faunal composition of the «cultural» beetles of Iceland and the Faeroes, emphasizing man as an «unpredicatable» agent of dispersal.

The Faeroese fauna of Carabidae comprises 26 species (Table 6.1) which is much fewer than found in the Shetland Islands (55 species according to Lindroth (1974) and Berry & Johnston (1980) although the two archipelagoes are of the same size 1400 km^2). Shetland and the Faeroes have 23 species (i.e. 89% of the Faeroese fauna) in common, and only *B. bipunctatum*, *B. geniculatum*, and *T. placidus* are not found in Shetland. On the other hand, the Shetland fauna includes several species of *Amara*, *Agonum*, *Pterostichus* plus entire genera not represented in the Faeroes. A comparison with Iceland, which covers an area about 75 times that of the Faeroes, shows that there are fewer species (22 according to Larsson & Gígja (1959) and Sigurjónsdóttir (1974)) than in the Faeroes and only 14 are in common (i.e. 54 % of the Faeroese). Thus, the most pronounced decrease in number of species occurs between Shetland and the Faeroes, but the difference in faunal composition appears to be greater between the Faeroes and Iceland. Why have not some of the non-Faeroese species of Carabidae, now occurring in Shetland or Iceland, or some of the common «ballast-species» of Lindroth (1957), managed to infiltrate the coexisting guild in the Faeroes? The size of the areas cannot explain the difference in species number between the Shetland Islands and the Faeroes, but it can be argued that the former are situated closer to the mainland and have a higher habitat diversity. Iceland is situated even further away from any large pool of potential colonizers, but has, on the other hand, a greater variety of habitats, apart from being much larger in size. However, with the high potential for dispersal by man that characterizes Ground Beetles (Lindroth 1957), distances may not cause any pronounced differences between the North Atlantic Islands. In the following I shall discuss the pattern of coexistence inferred from the Faeroese field-data in an attempt to point to factors which might determine the number of species in the islands.

The species diversity (and size-class diversity) is relatively low in all habitats (Table 6.2) and usually one (or a few) species completely dominate at a site numerically (Table 6.3; Fig. 6.3). The number of species, however, is similar in all habitats (Table 6.2). Environmental heterogeneity in space and time is generally assumed to allow more species to coexist when equilibrium prevails. The absence of woods and well-developed shrubs in the Faeroes reduces the possibility for a segregation in space, and climatic conditions and intensive sheep-grazing presumably increase uniformity of the landscape. Therefore, it is not surprising that practically all species of Carabidae are ground-living (*A. aulica* often climbes vegetation (Lindroth 1945: 107)). However, it is surprising that the species guild showed only a very moderate ($\alpha=0.37$) overall mean overlap in habitat utilization (Table 6.5), and that the species were well-spaced along the habitat dimension. Thus, it would seem possible for a few more species to invade the existing assemblages, provided that diffuse competition (MacArthur 1972) is not very strong. An increased number of species would seem even more likely when we take into consideration that the habitat classification, used in this study is coarse and that coexisting species of the family Carabidae probably often distinctly differ in their choice of microhabitat (e.g. Dawson 1965). A major objection to the arguments above is, of course, that this study only covers a short time and does not consider larval stages.

Temporal environmental variation is more difficult to measure and to assess. Seasonal and annual variations in climate are presumably relatively small, the summers are short, and there does not seem much scope for segregation in time. Although environmental conditions seem rather stable and predictable, reality is probably the opposite. Rare but recurrent spells of cold and warm weather, the very erratic daily weather situations, and strong human influence on the landscape suggest environmental instability and unpredicatability.

An overwhelming number of studies on syntopic species assume that competition is the key to the observed resource partioning (see Schoener 1974).

Faeroese Carabidae seem to partition the habitat dimension sufficiently well, and to be sufficiently dissimilar in size (implying differences in food ecology), to exist without excessively intense competition. In only six of 105 cases were two species so similar with respect to both dimensions (habitat and size) that a potential for competition seemed to exist. However, no signs of interactions such as changes in numbers or ecological shifts could be detected. The patterns of distributions and numbers do not deviate from what would be expected from randomness. Thus, on the basis of field-data, it is not possible to demonstrate the existence or absence of competition.

On islands, species often show ecological release which is frequently interpreted as a consequence of relaxed competition (MacArthur & Wilson 1967). It seems certain that several of the Faeroese Carabidae do have broader habitat niches than in Fennoscandia (from Lindroth 1945); this seems to be true especially with regard to *N. gyllenhali*, *N. salina*, *N. biguttatus*, *P. septentrionis*, *T. obtusus*, and *C. problematicus*. Thus, the pattern of coexistence among the Faeroese members of Carabidae can easily be interpreted to fit the theory of competition, although evidence of a competitive situation is wanting. Possibly the increase in species numbers with increased food-index (Fig. 6.2) indicates that food, in one form or another, is a limiting factor.

Then why are the species so well spaced along the habitat dimension? The simplest «explanations» would be that the field-data give a distorted picture, diffuse competition is strong, or other species of predators (e.g. spiders and other beetles) should also have been taken into account. However, in a thought-provoking paper, Wiens (1977) argues that in variable environments populations may pass through «ecological crunches» and relatively seldom reach a resource-defined equilibrium. After such bottle-necks, populations would experience less competitive or non-competitive situations with a surplus of resources that, according to theory of optimal foraging (e.g. Pyke et al. 1977), should lead to specialization and more clear-cut differences between species in resource utilization. Subsequent population growth would, however, soon restore a competitive environment. There is no unequivocal evidence for variable environmental conditions in the Faeroes, though it was argued above that prevailing weather conditions and human activities presumably lead to instability and unpredictability. Further support for such a view is given by the species-abundance curves (Fig. 6.3), which all fit a logarithmic series; this is consistent with what is usually found when environmental variability is great [see Stenseth 1979 for a recent review]. Under variable environmental conditions we would also expect the species to exhibit flexibility in their life strategies. The literature (West 1930, Lindroth 1945) does show that in fact 10 of the 15 common species (Table 6.4) overwinter as adults and larvae. The three *Pterostichus*-species, however, do overwinter only as adults, which is also true for *N. biguttatus* and *L. pilicornis* (though the latter may possibly overwinter also as larvae). Thus, circumstantial evidence points in the direction of environmental variability being important in determining patterns of coexistence and number of species of Faeroese Carabidae. This, of course, is not incompatible with the hypothesis that other factors (e.g. competition) also are important. It is interesting, and reasssuring, that Lindroth 1960, when discussing the relative poverty of the Carabid beetle fauna on the Azores compared with Madeira, emphasizes the importance of unfavourable climatic conditions; especially the wet summers.

Summary

Ground beetles were surveyed, by using pitfalls, sieving, and hand collecting, in 111 localities on 17 islands in the Faeroes. A total of 26 species are known from the Faeroes, 15 of which were common (wide inter-island distribution and/or high constancy). Number of species positively correlated with island size; species-area curve had a z-value of 0.25. Species diversity is low, habitat uniformity high, and mean number of species varied little between but considerably within habitat types. Despite large variation, there were significant positive correlations between number of species, species and size-class diversity, and an index of prey abundance. Within all habitat types species-abundance curves fitted a logarithmic series distribution. Although many species had broad habitat niches, overlaps in habitat utilization were usually low (mean $\alpha = 0.37$). Although species' sizes accord with the predictions of competition theory, no conclusive evidence of interspecific competition could be detected. The relative poverty of species and loose species-packing is discussed with reference to the presumably rather variable environmental conditions.

Acknowledgements

The late Professor Carl H. Lindroth was my teacher, colleague, close friend, and an inexhaustible source of inspiration. We shared a deep interest in the faunas of North Atlantic islands (Iceland in particular), and he encouraged the present work in the Faeroes. It was a great loss, that he did not live long enough to read and, no doubt, improve this paper. I wish to thank Drs P. H. Enckell, Lund, and T. Solhøy, Bergen, who were my co-workers in the Faeroes, Dorete Bloch Danielsen and J. K. Bærentsen, Tórshavn, and A. Fjellberg, K.-E. Erikstad, T. Klungland, Bergen, for valuable help in various ways. Stödisútbúgvingin (Academia Færoensis) provided laboratory facilities and the Royal Danish Navy assisted with helicopter transportation. The study was supported by grants from the Norwegian Research Council for the Sciences and Humanities (to S.-A. Bengtson) and the Nordic Council for Ecology (to T. Solhøy).

Reprinted with permission from Gustav Fischer Verlag, Stuttgart from: Entomologia Generalis 6 (2/4): 251–266; Stuttgart — New York: 1980 — 11. — ISSN: 0171–8177 EGMS — Nr. 270.

References

Bengtson, S.-A., & Hauge, E. (1979): Terrestrial invertebrates of the Faeroe Islands: I. Spiders (Araneae): Checklist, distribution, and habitats. – Fauna Norv., ser. B, 26 (2): 59–83, Oslo.

Berry, R. J., & Johnston, J. L. (1980): Natural History of Shetland. – Collins, London.

Brinck, P. (1966): Animal invasion of glacial and late glacial terrestrial environments in Scandinavia. — Oikos 17: 250–266; Copenhagen.

Connor, E. F., & McCoy, E. D. (1979): The statistics of biology of the species-area relationship. — Am. Nat. 113: 791–833; Chicago.

Dawson, N. (1965): A comparative study of the ecology of eight species of fenland Carabidae (Coleoptera). – J. Anim. Ecol. 34: 299–314; Oxford.

Hansen, K. (1966): Vascular plants in the Faroes. Horizontal and vertical distribution. – Dansk Botanisk Arkiv (issued by Danish Botanical Society) 24 (3): 1–141; Copenhagen.

Hespenheide, H. (1973): Ecological inference from morphological data. – Ann. Rev. Ecol. Syst. 4: 213–229; Palo Alto, California.

Hutchinson, G. E. (1959): Homage to Santa Rosalia or why are there so many kinds of animals? – Am. Nat. 93: 145–159; Chicago.

Jessen, K., & Rasmussen, R. (1922): En profil gennem en Tørvemose paa Færøerne. – Danmarks geologiske Undersøgelse IV: Bd. 1 (13); Copehhagen.

Jóhansen, J. (1971): A palaeobotanical study indicating a previking settlement in Tjørnuvík, Faeroe Islands. – Fródskaparrít (Annales Societatis Scientiarum Faeroensis) 19: 147–157; Tórshavn.

Larsson, S. G., & Gígja, F. (1959): Coleoptera 1. Synopsis of the species. – The Zoology of Iceland Vol. III (46a): 1–218; Copenhagen & Reykjavik.

Levins, R. (1968): Evolution in changing environments. – Princeton University Press, Princeton, New Jersey.

Lindroth, C. H. (1931): Die Insektenfauna Islands und ihre Probleme. – Zoologiska Bidrag från Uppsala Bd. 13: 105–599; Uppsala & Stockholm, Berlin.

Lindroth, C. H. (1932): Über die Geschichte der färöischen und isländischen Fauna. – Archiv für Naturgeschichte Bd. 1 (4): 673–675; Leipzig.

Lindroth, C. H. (1945): Die Fennoskandischen Carabidae. Eine Tiergeographische Studie. I. Spezieller Teil. – Göteborgs Kungl. Vetenskaps- og Vitterhets-Samhälles Handlingar VI, Ser. B, Bd. 4 (1): 1–709; Göteborg.

Lindroth, C. H. (1957): The faunal connections between Europe and North America.– Almquist & Wiksell, Stockholm & New York.

Lindroth, C. H. (1960): The ground-beetles of the Azores (Coleoptera: Carabidae) with some reflexions on over-seas dispersal. – Boletim do Museu Municipal do Funchal XIII, Art. 31: 5–48.

Lindroth, C. H. (1968): The Icelandic form of Carabus problematicus Hbst. (Col. Carabidae). A statistical treatment of subspecies. – Opusc. Ent. 33: 157–182; Lund.

Lindroth, C. H. (1974): Handbook for the identification of British Insects Vol. IV, Part 2: Coleoptera, Carabidae (issued by the Royal Entomological Society of London). – London.

MacArthur, R. H. (1972): Geographical ecology. Patterns in the distribution of species. – Harper & Row, Publishers, New York, Evanston, San Francisco, London.

MacArthur, R. H. & Wilson, E. O. (1967). The theory of island biogeography. – Princeton University Press, Princeton, New Jersey.

Pianka, E. R. (1974): Niche overlap and diffuse competition: – PNAS 71: 2141–2145.

Pianka, E. R. (1976): Competition and niche theory. – In May, R. M. (ed.): Theoretical ecology. Principles and applications: 114–141; W. B. Saunders Co., Philadelphia.

Pyke, G. H., Pulliam, H. R. & Charnov, E. L. (1977): Optimal foraging: a selective review of theory and tests. – Quart. Rev. Biol. 52: 137–154; State University of New York at Stony Brook.

Schoener, T. W. (1974): Resource partitioning in ecological communities. – Science 185: 27–39; Washington, D.C.

Sigurjónsdóttir, H. (1974): Hvenæfara skordýr og áttfætlur á kreik á vorin? – Náttúrufrædingurinn 44: 80–94; Reykjavik. (In Icelandic).

Stenseth, N. C. (1979): Where have all the species gone? On the nature of extinction and the Red Queen hypothesis. – Oikos 33: 196–227; Copenhagen.

Thiele, H.-U. (1977): Carabid beetles in their environments. A study on habitat selection by adaptations in physiology and behaviour. – Zoophysiology and Ecology 10: 1–369; Springer-Verlag, Berlin, Heidelberg, New York.

Turner, M. & Polis, G. A. (1979): Patterns of co-existence in a guild of raptorial spiders. – J. Anim. Ecol. 48: 509–520; Oxford.

West, A. (1930): Coleoptera. – Zoology of the Faeroes 40: 1–92; Copenhagen.

Wiens, J. A. (1977): On competition and variable environments. – American Scientist 65: 590–597; Washington, D.C.

Wilson, D. S. (1975): The adequancy of body size as a niche difference. – Am. Nat. 109: 769–784; Chicago.

Author's address:
Prof. Fil. dr. S.-A. Bengtson
Museum of Zoology
University of Lund
S-22362 Lund
Sweden

7. The land systems of the Faeroe Islands

G. K. Rutherford & C. E. B. Taylor

The land survey of the Faeroe Islands involved four levels of intensity: the land region, land district, land system and land type.

The Land Region is termed the Faeroe Island Tertiary Plateau Basalt Land Region. The entire region reflects strongly glaciated landscapes developed in Tertiary Plateau Basalts that now support a vegetation indicative of the cool temperature oceanic climate.

Within the Land Region, two Land Districts are defined. The strongly glaciated, NW-SE trending islands of dominantly Upper Basalt Series are characteristic of the Northern Land District. The Southern Land District is characterized by strongly glaciated islands of predominantly Middle and Lower Series basalts and intrusions.

The Land Systems are subdivisions of the Land Districts characterized by a recurring pattern of landforms, soils and vegetation.

Each Land System is illustrated and in turn is divided into smaller, more homogeneous Land Types. The Land Type is recognized as an area of land on a particular parent material, and has a fairly homogeneous combination of soil and chronosequence of vegetation.

Eight land systems are described on the Faeroe Islands. On the basis of broad similarities these land systems can be grouped into three categories.

The mountain land systems (Kunoy, Videreidi) are dominant in the northern islands, covering roughly 35 % of the total area of the Faeroes. Characterized by bedrock outcrops, Regosols and Podzols, these land systems have little or no vegetation.

The plateau land systems (Torshavn, Kaldsbakfjord) represent 30 % of the total area. The land systems dominate areas of Upper Basalt Series.

The valley land systems (Saksen, Sandvik, Hvalbøur, Våg) predominate the Middle and Lower Basalt Series. The valleys are observed to have moderate to steep backslopes and gently undulating to level valley floors.

Each land system is represented by a vertical cross-section and a tabular description of the landforms and soils associated with the land system. The illustrated cross-sections were selected so as to provide an indication of the general character of the land system and of the relationship of one land type to another.

Monographiae Biologicae, Vol. 46 ed. by G. K. Rutherford

Torshavn land system

Land type	Landform(s)	Soils
TA	Bedrock, rockfalls and rubble colluvial aprons (slopes of 30–45°) progress downslope to form a thin blanket of colluvium over cirque till material (5–10°). Outcrop bedrock to 2 m height. Local waterlaid deposits to 1.5 m thickness.	Fine textured Regosols and thick Regosols. Strongly eroded Regosols develop into deep basin peats and Gleysols. Ochreous B horizons frequent. Extensive areas of Brunisol soils.
TB	Strongly eroded antiplanation terraces. Planar with low slope. Local outcroppings to 1.5 m height. Deflation and gully erosion. Open lag gravels, not stoney and block fields. Solifluction terraces and sorted pattern ground (slopes 0.5–8°).	Soils vary in thickness from bedrock (10 cm) to 2 m thickness. Strongly disturbed, eroded fine textured Brunisols and peaty soils on local sandy alluvium. Brunisols with sorted patterned ground.
TC	Steepland till over slopes from 5–35°. Deflated and strongly gullied. Erosion of basalt layers produces sandy solum and large blocky rubble. Mass wasting and sheet erosion of surface soils.	Steep upland slopes (30°) contain large (1–2 m) blocky rubble. Thin Podzols and basin peats. Colluvial-water laid material forms peat or thick undifferentiated soils in local depressions buried peat and organic horizons. Sheep overgrazing and peat removal hasten erosion.

Plate 7.1. Torshavn Land System illustrating the TB and TC land types and strong erosion of the land surface.

TD	Alluvial over thick dense angular basal till. Hummocky with deeply eroded channels. Microrelief is 2 m.	Thick, partly to strongly decomposed peat deposits. Local development of thin peat podzol with a narrow, irregular-broken iron pan over saprolite.
TE	Hummocky dense till deposited over low concavo-convex slopes. Very stoney surface and deeply gullied, strongly deflated and eroded peat extraction areas.	Local thick aeolian peat deposits, gley soils and poorly decomposed stoneless peat soils. Extensive areas of peat extraction.

Kaldsbak-Kollafjord land system

Land type	Landform(s)	Soils
KE	Bedrock, horns and arêtes of upper basalt. Very steeply sloped (15–30°) with thick blanket of talus and scree. Veneer of aeolian silts. Col areas (0–7°) reveal deep blanket of steepland till overlain by waterlaid materials and aeolian silts. Lag gravels and deflation manifest when depth to bedrock <1 m. Blockfields and frost weathering predominate.	Bedrock, sandy screes over fine, undifferentiated dark brown material to depths of 2 m. Dark organically stained surface horizons and some development of ochreous B horizons. No large stones, sorted patterned ground on slopes of 2–8° and depth to bedrock <2 m.

Plate 7.2. Kaldsbak-Kollefjord Land System. View to the NW from a col between Kaldsbak and Kollefjord, at an altitude of 295 km.

Torshavn Land System

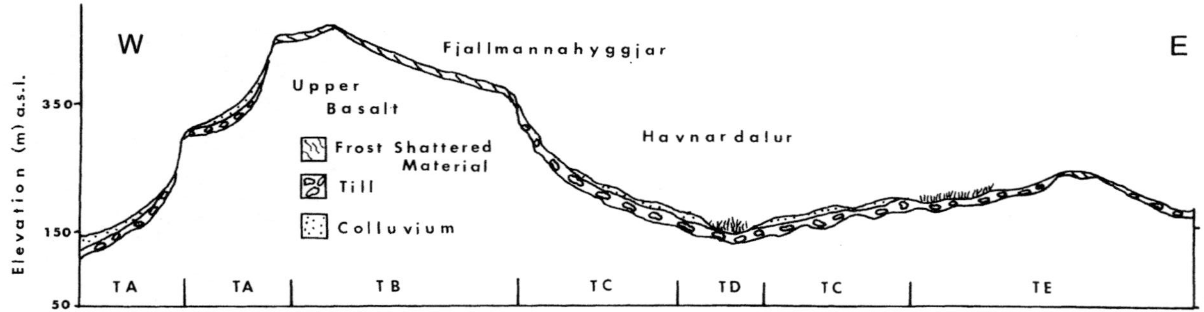

Fig. 7.1a.

Fig. 7.1b.

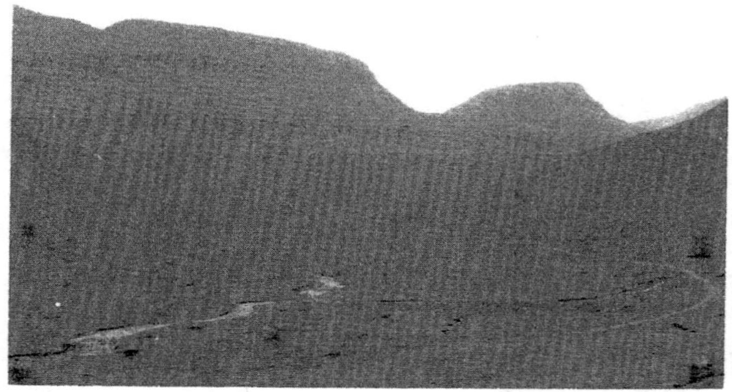

Plate 7.3. View of typical Torshavn Land System. Note removal of a large amount of surface material due to sheet erosion.

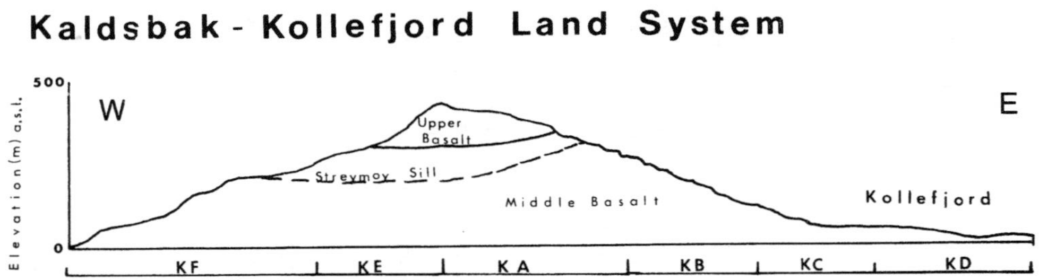

Fig. 7.2a.

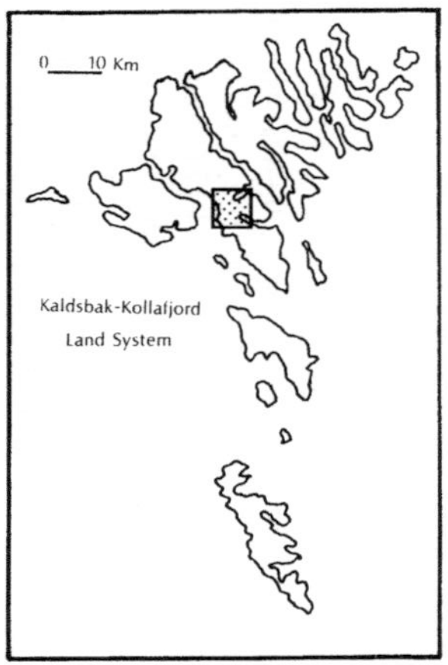

Fig. 7.2b.

Plate 7.4. Lag gravels and severe wind erosion on the Streymoy sill in the KE land type.

KF	Steepland and basal till overlain by steepland colluvium and thin layer of waterlaid materials. Very hummocky with deeply eroded and gullied terraces (15–25° slopes). Local outcrops and *hamnars* are 2 m height and provide steep (30°) vegetated scree aprons and narrow terraces (2–8°).	Regosols and undifferentiated Brunisols, deflation peats, gleyed Brunisols, and Podzols. Numerous organic laminae, thin buried organic horizons, disturbed horizons and solifluction deposits.
KA	Steeply sloped arêtes, peaks and hamnars (60°) covered in deep deposits of steepland till and colluvium and rubbly colluvium (45° slopes). Wind deflation, frost action and fluvial processes produce thick (2 m) deposits of fine textured material on adjacent terraces. Stone lines and solifluction terracing. Local depressions collect water laid material to 3 m depth.	Bedrock, very sandy scree (30 cm) overlying finer undifferentiated material. Some mottling with depth, stone free Regosols and Brunisols.
KB	Long steep concave slopes (25–30°) and terraces broken by hamnars (5–6 m height) overlain by very deep steepland till, colluvium and water laid materials. Undifferentiated and mass wasting, erosion dominate. Large angular-blocky rubble common on surface. Terraces often exhibit rock-knob lakes (0–6°). Terraces have adjacent hamnars of 2–3 m height.	Numerous buried organic horizons and fine gravel layers (Tjornvik-Saksen Complex). Ferric horizons dominate but no induration or orstein. Podzols develop adjacent to slopes and deep aeolian peats in local lowlying areas.
KC	Base of long concave slopes (5–15°) covered in deep steepland till deposit, steepland colluvium and waterlaid materials. Very hummocky with complex microtopography. Till forms resemble glacial mounds (roches mountonée, drumlinoid). Strongly gullied and eroded.	Thick Podzols developing in moderately well drained areas. Soils often degraded or contorted by solifluction and mass wasting. Bright ferric mottling with depth (50 cm).
KD	Recent alluvium overlying dense bouldery clay and basal till material. Level to planar. Some roche mountonées evident.	Undifferentiated alluvium and highly organic soils. Moderate-imperfect drainage to 70 cm then very strongly gleyed.

Saksen land system

Land type	Landform(s)	Soils
SkA	Steep complex slopes (30–40°) overlain by steepland colluvium. Gully and sheet erosion. Represents the sides of a pre-existing glacial 'U' shaped valley. Occasional resistant knobs or peaks or upper basalt; prominent cirques in middle basalt covered in cirque till and steepland covered in cirque till and steepland colluvium. Inclusions of basal till, erratics. Major river valleys are cut in flat bottom of pre-existing glacial valley.	Bedrock and angular coarse sandy screes. Regosols eroded and deflated; shallow profile (30 cm) mottled at depth; progressively thicker downslope. Several thick (3 cm) buried organic horizons encounter due to slumping and/or terracettes. Cirque basins are wet, partly decomposed peats and Gleysols, mottled at depth. Inclusion of sandy material due to piping on terraces.

SkB	Long simple slopes (20–25°) of middle basalt covered in cirque and steepland till and colluvium. Deep, wide gullies prevalent. Very thick colluvial deposits (10 m). Slumping and mass wasting.	Poorly developed Podzol soils orstein but strong thick ochreous horizons. Distinct mottling in B horizons.
SkC	Dense, thick, angular basal till. Strong undulation (5–15°) overlain by alluvium in many areas. Local thin till layers overlie glacially smoothed bedrock.	Thick peat deposits juxtaposed to Podzol. On stable, imperfectly drained toe slope sites, soils are organic with thick peat horizons alternating with thick mineral horizons (Tjornuvik-Saksen Complex).
SkD	Recent thick alluvium over dense bouldery clay and basal till. Till deposits are very hummocky and microrelief is complex. Deeply dissected and disturbed by removal of peats.	Ground water gley soils and thick (1 m) (Saksen type) peat. Inclusions of very thin-thick organic soil association. Sandy loam A horizons over peat.

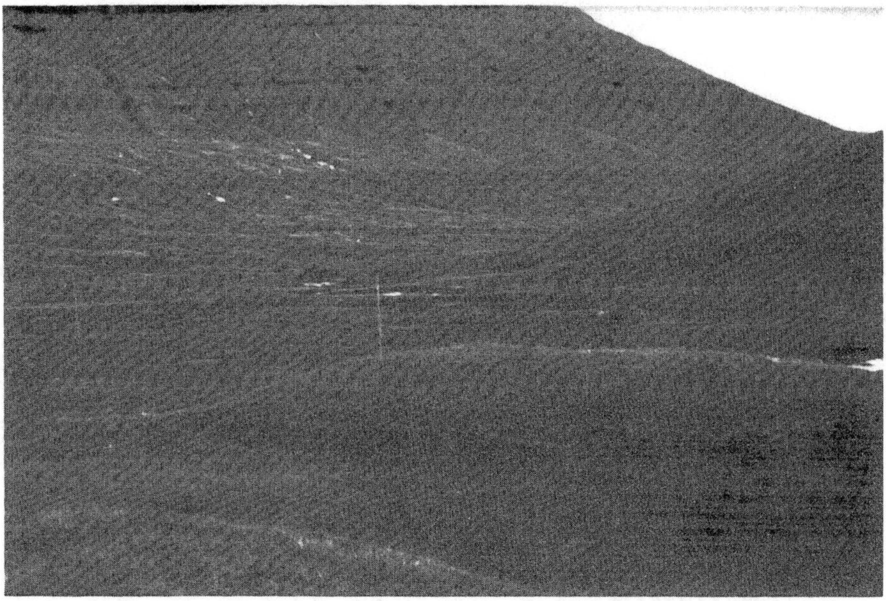

Plate 7.5. Saksen Land System viewed to the NE. The SkD land type: Organic soils developed on a dense bouldery basal till.

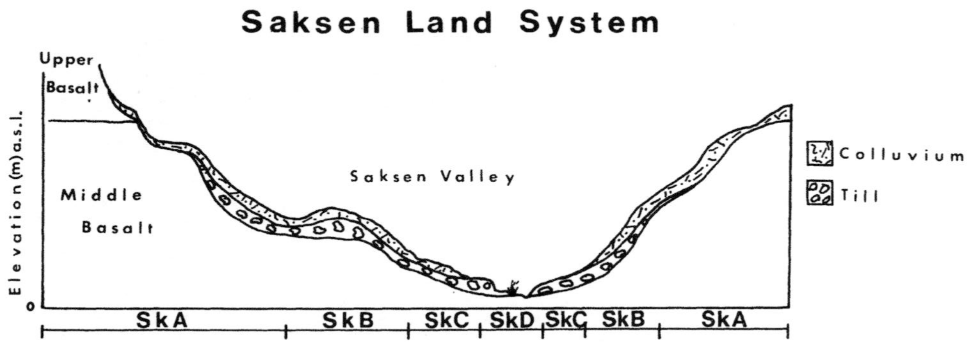

Fig. 7.3a.

Fig. 7.3b.

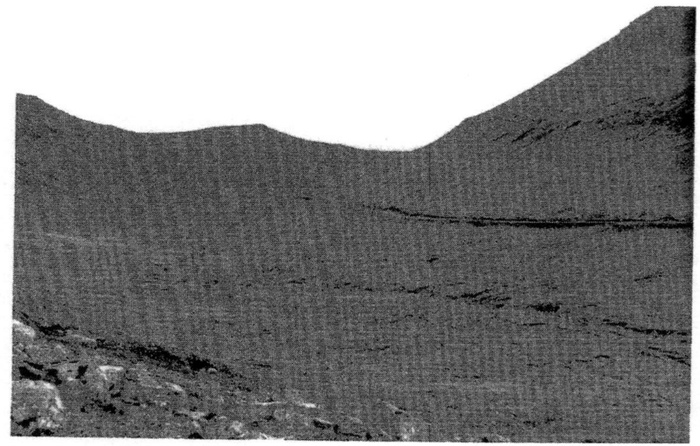

Plate 7.6. An eide, as seen looking north on the road from the Slaettintindur basin.

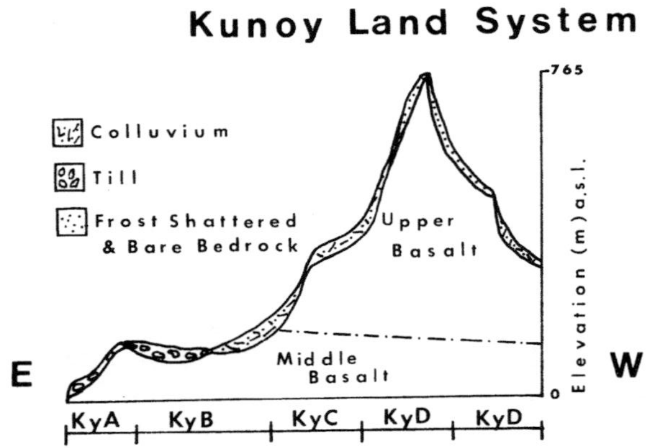

Fig. 7.4a.

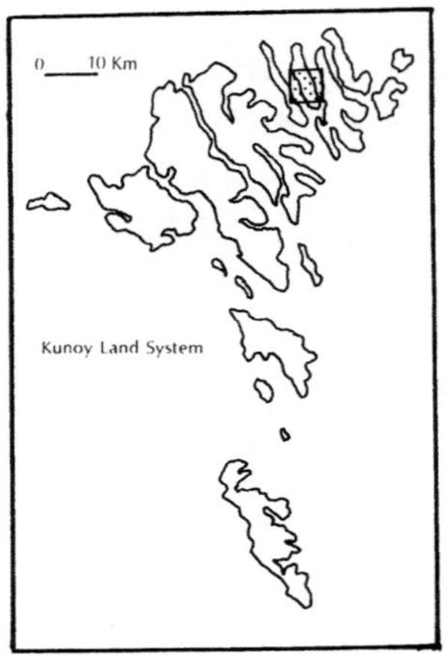

Fig. 7.4b.

Plate 7.7. The KyC land type is dominated by Organic and Podzolic soils.

Kunoy land system

Land type	Landform(s)	Soils
Ky_D	Very steeply sloping thin layers of upper basalt covered in fine textured talus and sandy colluvium. Nivation hollows, strongly eroded and dissected. Altiplanation surfaces often present covered in blockfields and frost shattered material.	Bedrock and open non-vegetated lag gravels, scree piles and blockfields. Lichen and mosses provide A horizon. Few thin Regosols.
Ky_C	Cirque form with steeply sided (30–45°) headwall and several narrow benches. Coarse angular cirque till overlain by very blocky rubble and steepland colluvium.	Non-vegetated Regosols; local deep undifferentiated sandy loam soils in association with finely textured thin Regosols. Solifluction terraces and buried thick organic horizons develop into thin basin peats (often degraded) and Podzols.
Ky_B	Basin of cirque dominated by steepland till and colluvium overlaid by water laid materials. Large blocky erratics. Slopes are short and range from 2–25°.	Deep partially decomposed basin peats (Saksen type) and Podzols.
Ky_A	Thick steepland colluvium overlying angular steepland till. Very deeply gullied and eroded. Slopes are short and range from 0–30°.	Podzols with thin iron pan. Saprolite and indurated till.

Plate 7.8. The Kunoy Land System, illustrating a deep cirque carved into the Upper Basalt, is indicative of the northern islands.

Videreidi land system

Land type	Landform(s)	Soils
VA	Cirque and nivation hollows in thin layers of upper basalt. Very steep slopes (30–55°) covered in fine textured talus and scree cones. Deeply eroded and gullied.	Bedrock and coarse angular sandy screes overlain by thin layers of fluvially deposited sands and silts.
VB	Thin angular cirque till and nivation debris overlying dense steepland colluvium. Water laid material in strongly undulating basin areas. Strongly gullied and eroded.	Regosols and sandy loam Brunisols with high organic matter contents. Partially decomposed basin peats and gleyed peat soils. Inclusions of thick alternating layers of recent alluvial matter burying peaty horizons to a depth of +50 cm.
VC	Strongly eroded and gullied slopes of 15–25°; composed of thick, dense basal till overlain by hummocky to strongly undulating thin layer of steepland colluvium.	Strongly decomposed basin peat and Podzols with thin iron pan. Stony surface horizons. Some bog iron ore and gley due to local depression or dense till with permeable ortstein. Strong weathering of till (saprolite).
VD	Thick, compact bouldery till overlain by coarse alluvium. Undulating to smooth simple slopes (2–10°).	Thin stoney A horizons and *bøur* cultivation. Peat is excavated and reveals bog iron ore. Podzols show strong chemical weathering of till material

Plate 7.9. View of Videreidi Land System with typical steeply sloping eidi walls.

Videreidi Land System

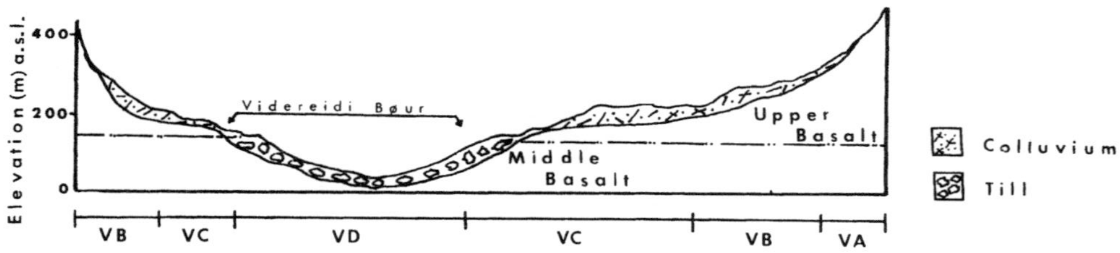

Fig. 7.5a.

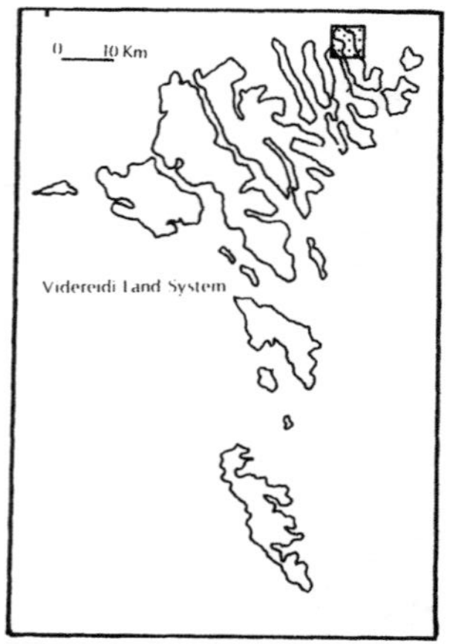

Fig. 7.5b.

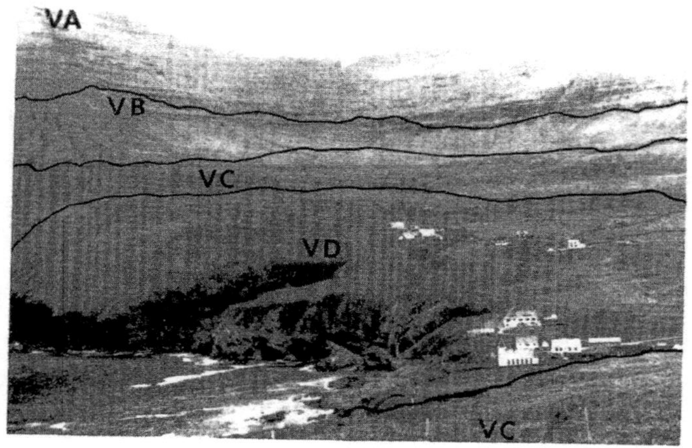

Plate 7.10. Land types of the Videidi Land System: Videreidi, Vidoy.

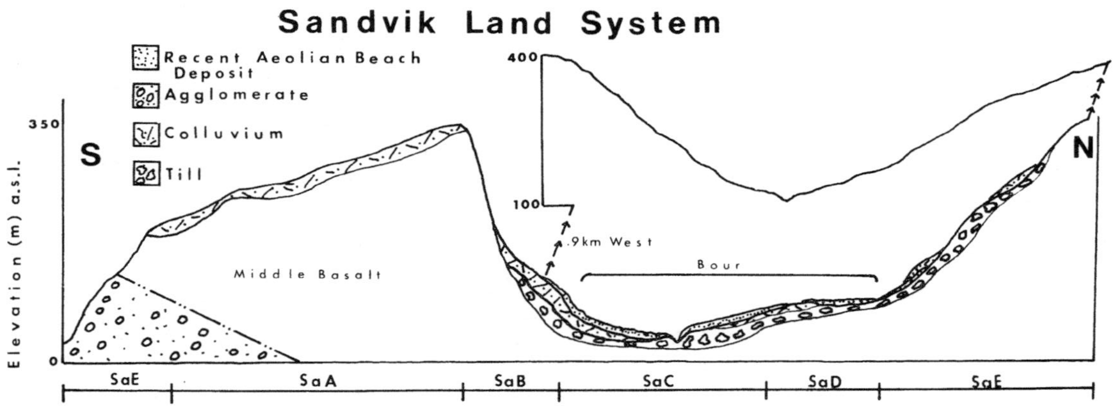

Fig. 7.6a.

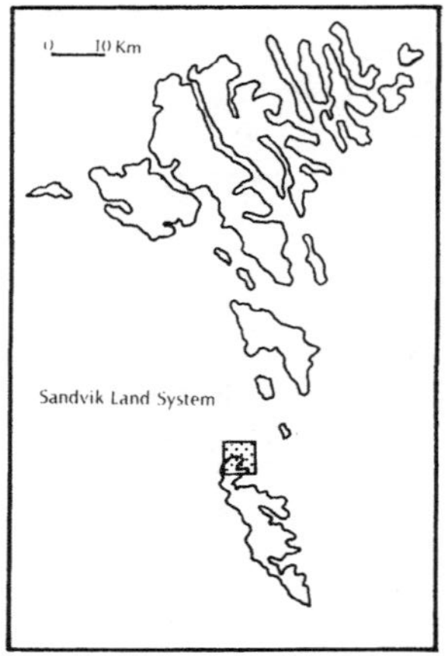

Fig. 7.6b.

Plate 7.11 View to the east from Sandvik illustrating land types SaD, SaE.

Sandvik land system

Land type	Landform(s)	Soils
SaA	Altiplanation surfaces, lag gravels and blockfields. Planar to slightly sloping (1–5°). Local outcrops of 2 m height and erosion areas by aeolian and fluvial agents. Inclusions of steepland colluvium.	Sandy-silt Regosols over frost shattered debris of 55 cm depth. Local thin deflation peats with inclusions of very sandy alluvial material. Strong deflation and sheet erosion. Some solifluction of Brunisolic soils.
SaB	Steeply sloping (30–45°) eidi valley walls with steep hamnar. Deeply gullied, slumped and steepland till covered by aeolian beach sands 1 m depth on 30° slopes adjacent to beach and coastal areas.	Finely textured stoneless Brunisols which are strongly disturbed and contorted. A horizons are often buried due to slumping of material. Sandy surface horizons to depth of 2 m overlying slightly weathered till material.
SaC	Eidi valley floor composed of hummocky compact basal till and colluvial material of variable thickness (1.5–3 m). Slopes are long and range from 2–8°. Recent marine sands cover much of area adjacent to coast to a depth of 2 m.	Dark brown sandy loam A horizons overlying very thin partly decomposed peat layer and distinct brightly coloured B horizons. Till and colluvial material is just slightly weathered, but is very dense and bouldery.

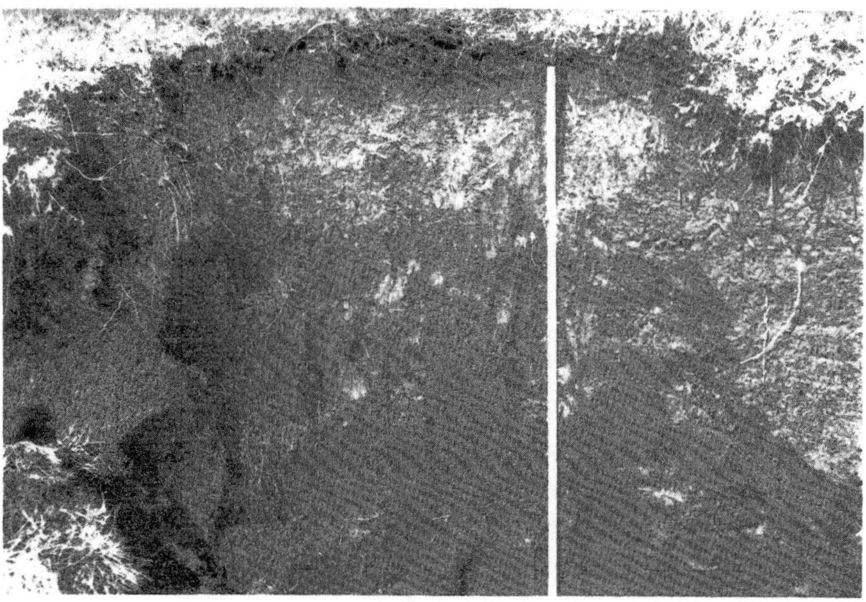

Plate 7.12. A Brunisol, typical of the Sandvik Land System.

SaD	Plateau and terrace of eidi floor comprised of thick, aeolian material overlying a compact basal with minor inclusions of steepland colluvium; slopes range from 2–10°. Depth to bedrock is variable (40 cm–2.5 m) over hummocky terrace area and numerous outcroppings and rock knob lakes are present.	Thin sandy soils with gravelly loam B horizons often overlying a variable thickness of peal or bedrock. Inclusions of deep, aeolian peats with sandy-silt laminae.
SaE	Steep slopes of eidi valley covered with steepland till and local, thin deposits of colluvial material.	Thin peaty mineral soils on coarse gravels which, when peat is removed, are severely eroded by wind and water. Col or open, flat areas consist of deep lag gravels (75 cm) and Brunisol soil forms.

Hvalbøur land system

Land region: Faeroe Islands Tertiary Plateau Basalts.

Land district: Suduroyar.

Geology: Middle basalt series coal bearing sequence.Lower basalt series.

Geomorphology: A very steep sided glacial U-shaped valley eroded through the Middle and Lower basalt series. Severe mass wasting, and wind deflation on the antiplanation surface has produced very large lobate colluvial slopes and vertical hamnar on the Middle basalts. This material then contributes to the thick colluvial and steepland till material that covers the Lower basalt slopes. The valley floor is flat and covered in a thick deposit of alluvium. Relief ranges from sea level to 450 m and slopes are generally greater than 25°.

Land type	Landform(s)	Soils
HA	Planar antiplanation surface characterized by thin soils, blockfields and lag gravels, solifluction deposits (2–8°). Strongly deflated and sheet erosion of land surface.	Sandy thin Regosols over frost shattered debris. Local thin deflation peats with inclusions of very coarse sandy material. Strongly deflated and undifferentiated Brunisols.
HB	Upper slopes (45°) are exposed bedrock and thin screes. Uncharacteristic Middle basalt landscape with steep cathedral-like tower shaped nivation hollows. Thick rubbly colluvial material progresses to a thick (2–3 m) hummocky blanket of angular steepland colluvium. Terraces formed by hamnar and are strongly dissected.	Undifferentiated Regosols very coarse becoming thicker with length of slope. Partly decomposed organic peats with silty A horizons develop in local depressions adjacent to terraces. Till material and steepland colluvium is slightly weathered, gullied and eroded. Podzols with irregular, broken or degraded thin iron pans.
HC	Thick alluvium overlying dense heterogeneous bouldery basal till.	Undifferentiated alluvium and sandy organic soils with inclusions of Gleysol soils.
HD	Rough, highly dissected plateau with outcrops of 2 m height. Strong erosion and sheetwash. Frost action very prevalent.	Deep, undifferentiated peat soils develop from aeolian material over coarse sandy alluvium. Strongly deflated and eroded lag gravels and coarse angular debris.

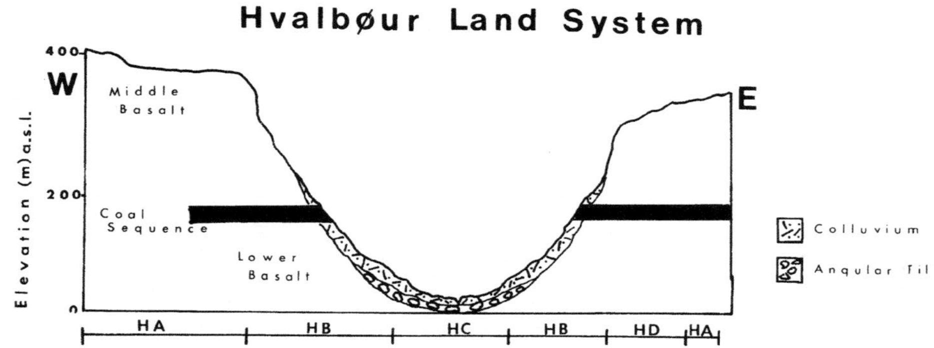

Fig. 7.7a.

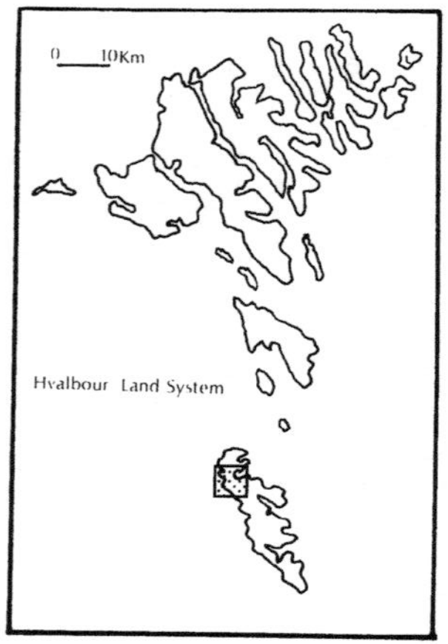

Fig. 7.7b.

Plate 7.13. View towards Orgnafjordur from col illustrating HA and HB land types.

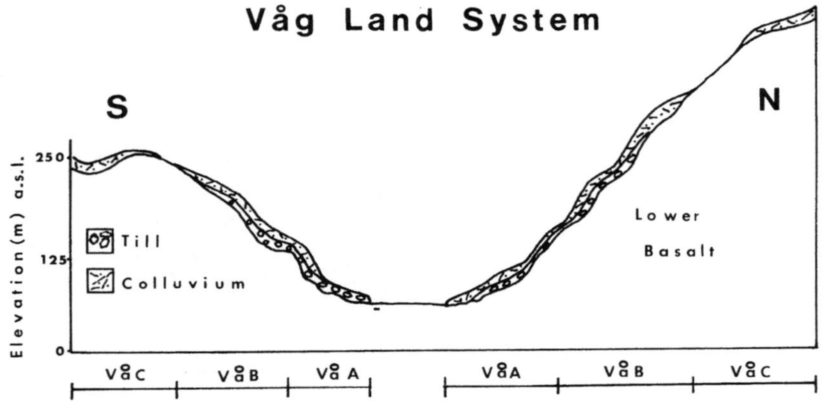

Fig. 7.8a.

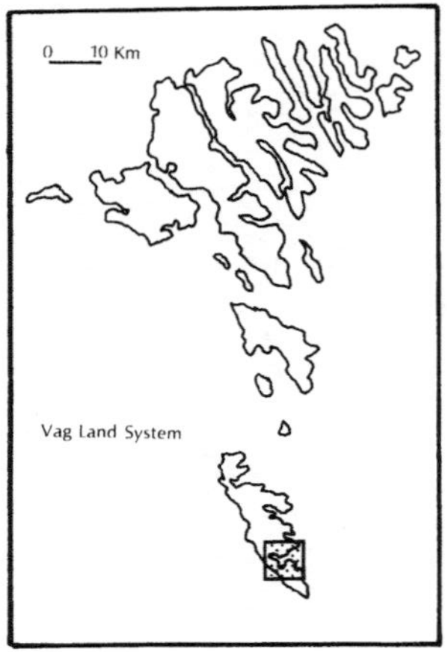

Fig. 7.8b.

Plate 7.14. Thin layer of dense, gravelly Steepland till overlying glacially smoothed bedrock surface and development of a Podzol.

Våg land system

Land type	Landform(s)	Soils
VåA	Valley walls of eidi and fjord have steep slopes (20–25°) and are broken by gently sloping, broad terraces (.25 km wide). Terraces are covered with a thin angular basal till deposit. Till is very dense and indurated. Discontinuous ortstein development. Coarse, hard, angular debris forms shallow, short, rubbly debris slopes from eroded hamnars (2 m) of Lower basalt.	Shallow Regosols. Surface horizons are characterized by coarse texture, and stoney angular fragments in thin surface horizons. Podzols are developed on an indurated, dense glacial till and thin angular colluvium. Podzols with strongly developed iron pan (2–5 cm) and occur from 25 cm to 55 cm in the profile form. Imperfectly drained.
VåB	Large, very broad upland valley slopes (.5 to 1 km wide; 15–30°). Covered with a thinner deposit of basal (min. 20) till and steepland colluvium. Debris on slopes or talus from outcrops of lower basalts is very sharply angular, not vegetated. Slumping and mass movement of slopes very evident.	Very thin gravelly surface horizons developed over partly decomposed peat. Podzols with thin albic horizon evident on dried profile face. Organic soils can develop 60 cm thick peat horizon due to local depressions in bedrock. Removal of soil (peat) and subsequent erosion causes large bedrock or gravelly scars.

Plate 7.15. Very coarse, angular debris and shallow soil development, showing VåA land type.

VåC	Upland plateau (5–10°). Hummocky broad, concave slopes and basins broken by narrow (1–2 m) hamnars. Scree slopes are composed of coarse angular fragments of lower basalt. No evidence of glacially compacted till material; minor inclusions of steepland colluvium.	Thin Podzols primarily developed over glacially smoothed bedrock. Gleysols evident. Soils are strongly eroded by wind and water. Lag gravels and Brunisols do not form polygonal ground.

Authors' addresses:
Dr G. K. Rutherford
Dept. of Geography,
Queen's University,
Kingston, Ontario,
Canada.

Dr C. E. B. Taylor
Ontario Hydro,
Toronto, Ontario,
Canada.

8. Soils of the Faeroe Islands

G. K. Rutherford & C. E. B. Taylor

Although the Faeroe Islands lie on the same latitude as Western Norway and Central Scandinavia, the soil forming conditions are markedly different. The soil environments in Central Scandinavia, with their longer and warmer summer climates, support boreal forests and a much greater range of vegetation; two of the soil forming factors which have made major contributions to the greater diversity of soil types. In the Faeroe Islands the high and almost continuous precipitation, the low annual temperature, the steep and rugged terrain and the lack of boreal forest mitigate against a wide diversity in the soil pattern. In spite of these limitations, five clearly defined soil types may be determined in the field.

The soils generally have much deeper organic horizons than similar soils reported in more southern latitudes; some deeper organic horizons are composed of strongly humidified to dark amorphous materials. Most soils are continuously wet and deep peat surface horizons are formed in poorly drained sites up to 800 m a.s.l. with the exception of bright coloured iron pan horizons, strong colour development and marked horizon differentiation are uncommon. Strong textural differentiation is common place in soil profiles but this is of sedimentological and not pedological origin.

The mineral soils of the Faeroe Islands are friable and composed of predominately silt-sized materials (Tables 8.1 & 8.2). Loss on ignition values range from 6 % in a C horizon to 98 % in a buried humic layer. Surface horizons are strongly acidic and the profiles become less acidic with depth; horizons with higher loss on ignition are always more acidic. The cation exchange capacity ranges from 23 meq in Regosols to 82 meq in Organic soils. The average base saturation is about 20 % and the exchange complex is dominated by H but Ca and Mg are also significant. Extremely high values for Na are reported in several profiles particularly in windy sites near coasts. Although in this archipelago all sites are less than 3 km from a coast.

The description and grouping of the Faeroe Island soils is based on profile descriptions and have been classified to Great Soil level using the Canadian system of soil classification (Canadian Soil Survey Committe 1978). It has been

Monographiae Biologicae, Vol. 46 ed. by G. K. Rutherford

Table 8.1. Morphological data from some soil profiles on the Faeroe Islands. Abbreviations: Canadian Soil Survey Committee 1978; (H) Von Post and Granlund 1926.

Horizon	Depth cm	Colour	Texture		Structure/Consistency
Regosols					
Of+Om	8–0	10 YR 2/1			
A	0–19	10 YR 2/1	sl		
Bh	19–40	10 YR 2/1	sl		
Cc	40–62	10 YR 5/4	fsl	c	3bk; mfi, cw, sir
Brunisols					
Of	10–8				(H4)
Om	8–3	5 YR 2/1	s		(H5)
Oh	3–0	10 YR 3/1	s	m	2sbk; wso
Al	0–43	10 YR 3/1	sl	c	2bk; ws
Bh	43–108	10 YR 3/2	sil		
Podzols					
Of	10–80				
Om	8–3	7.5 YR 4/2	fs		(H2)
Oh	3–0	7.5 YR 4/4	sl	c	2b1k; (H6)
Al	0–20	7.5 YR 4/3	qsl	f	3sbl; sbk; wps-mfr
Aej	20–35	10 YR 4/3	ls	f	2bk; wso-mfr
Bli	35–36	2.5 YR 4/8	consir		
L2c	36–87	10 YR 4/3	gls	c	2sbk; mvfr, cw
B3	87–120	10 YR 5/4	gls	c	2sbk; mfr
Cc	120–150	7.5 YR 5/4	gls	c	2m; mfi, cs
Gleysols					
Of	2–0				
A	0–6	10 YR 3/1	sl	f	lsbk; wss-wp
Ag	6–18	10 YR 3/1	sl	m	2sbk;
Bl	18–26	7.5 YR 5/6 (w)	ls	m	2sbk; ws-mfi
B2g	26–55	2.5 YR 6/2 (N4/0)	sl	m	3bk
B3g	55–69	5 YR 3/1	sil	c	2m; mfr
C	69–	10 YR 4/3	sl	vf	wsp-mfi
Organic soils					
Aj	0–22	10 YR 2/2	sil	f	lsg; wso-mfr
Of	22–40	2.5 YR 2/2			; wss (H4)
Om	40–55	2.5 YR N2/0			; ws (H5)
oH	55–117	2.5 YR 3/2			(5–6)
Cj	117–	10 YR 3/2	asil	m	3sbk; ws

Table 8.2. Results of some physical and chemical analyses of Faeroe Island soils.

Horizon	Depth cm	*Particle-size (%)			Organic Matter (%)		pH 0.01 n	*Exchange Cations meq/100 g soil 2mm				CEC	BS %
		sa	si	cl	C	LoI	$CaCl_2$	Na	K	Ca	Mg		
Regosolic soils*													
Om	8–0	45	52	3		47.0	4.8	1.8	1.1	1.5	0.8	58.0	9
A	0–19	48	47	5	8.7	25.5	5.2	1.3	0.7	1.3	0.6	56.0	7
C1h	19–32					38.7	5.0	1.1	0.4	1.5	0.9	57.9	7
C2c	40–50	55	43	2	6.5	26.3	5.3	0.7	0.4	1.1	1.6	33.2	11
C2c	50–62	53	43	4	3.4	12.2	5.5	1.7	1.0	1.3	0.3	30.6	14
Brunisolic soils													
Oh	4–0	65	35	0	5.2	27.1	5.16	3.9	1.2	3.6	1.5	35.2	29
Al	0–20	72	28	0	5.4	31.9	5.39	3.6	1.3	4.8	3.3	39.8	33
	20–43	21	79	0		46.9	4.87	3.4	1.3	3.9	3.9	47.5	26
Bh	43–75	43	54	0	4.6	27.5	4.55	6.7	1.3	4.8	2.0	38.3	39
	75–95	72	25	3	3.7	22.5	3.74	3.3	1.3	3.9	3.3	41.7	28
	95–108	86	13	0	4.2	22.1	4.24	4.5	1.0	2.9	1.4	30.7	32
Podzolic soils													
Al	0–20	51	42	7		11.6	4.4	1.9	0.7	1.1	1.9	35.4	15
Aej	20–35	55	41	4	2.0	6.6	4.9	1.1	0.6	1.2	1.4	21.7	26
Bli	35–36	66	32	2	0.3	5.1	5.1	2.1	1.1	8.9	7.7	29.3	22
B2c	36–87	76	24	0	0.4	5.3	5.3	4.4	2.2	4.5	6.6	35.5	56
B3c	87–120	79	21	0	0.2	6.5	5.5	1.2	0.6	10.0		38.3	
Cc	120–150	85	15	0	0.0	5.4	5.5	2.1	0.6	19.5			
Gleysolic soils													
A	4–0	43	51	6	10.5	19.3	4.7	2.8	1.2	3.9	6.9	38.8	38
Ag	0–17				6.7	17.0	4.7	2.3	0.9	2.2	3.5	36.8	25
B1g	17–26	51	41	8	3.3	11.5	4.9	1.2	0.7	1.1	1.5	26.0	19
B2g	26–55	48	47	5		33.9	5.1	1.7	0.3	1.0	1.8	25.9	19
B3g	55–69	48	46	8		25.8	4.9	2.2	0.5	1.2	2.0	34.9	17
C	69–	52	45	3		12.8	5.0	2.0	0.7	1.7	2.0	28.4	23
Organic soils													
	0–22					36.2	4.6	1.5	.5	2.6	.8	58.5	9
	22–40	H4				63.3	4.2	1.1	.2	1.5	.5	59.2	5
	40–55	H5				80.1	4.5	1.7	.1	4.2	.9	81.7	24
	55–117	H6				98.5	4.5	1.8	.1	5.0	.9	80.5	23

*As many soils, and in particular Regosolic soils, may contain high amounts of gravel it must be emphasized that these results will greatly exaggerate whole soil values.

found possible to indicate that there may be more than one general type in a Great Soil Group. No effort has been made to classify the soils further in Great Subgroup or Family, but sufficient field and laboratory evidence exists to indicate that certain variations over the central theme may occur.

Regosols

Regosols in the Faeroe Islands generally occur as ungleyed shallow mineral soils with a very weakly developed subsoil horizon. In some locations a thin black coloured topsoil and weak B horizon development is observed and is attributable to mottling and gleying caused by excessive moisture conditions. The surface horizons are extremely acidic in reaction and become very strongly to strongly acidic with depth. Regosols most frequently have blackish sandy to gravelly loam surface textures and the surface horizons are very high ($<30\%$) in organic matter. The depth of the Regosols ranges from 4 cm overlying outcrop bedrock (bald scars) to over 60 cm depth in areas of aeolian deposition.

Regosols occur in the Faeroes on the altiplanation surfaces, plateau and eroded col areas of the Upper Basalt series and are derived primarily from shallow deposits of glacial drift, blockfields, head deposits and the products of mass wasting. They are frequently found on unstable sites which are subject to severe frost action and wind erosion. The vegetation communities common to the Regosols are similar to those found in most alpine and high mountainous grasslands. Much of the areas of Regosol development are devoid of vegetation as a result of the removal of peat, overgrazing by sheep, unstable site conditions, violent wind and fluvial erosion or a combination of any of these factors. *Rhacomitrium langmosium* (Wooly Hair Moss) and lichens are the first colonizers and humus forms on newly stabilized rock outcrops, debris or slope deposits (Pennington 1970). Upon this stabilized carpet, grasses such as *Festuca ovina*, *F. vivipana*, *F. rubra* and *Poa pratensis*, are established, often in close association with *Calluna vulgaris* and *Juniperas communis*. Mat grass (*Nardus stricta*) occurs in many areas where overgrazing by sheep has tended to suppress the previously mentioned grasses. When a thin A_0 horizon is present it is observed that the root material is well preserved, and this is most likely attributable to limited microbial activity due to the lower temperatures observed at Regosol sites.

The exchange capacities of the surface A horizons of Regosol soils range from 49.8 meq/100 g to 55.9 meq/100 g. These relatively high exchange values are attributable to the very high organic matter content of these horizons as expressed by a black or dark brown colour and LoI values in excess of 22%.

The subsoils are sticky when moist, friable when dry and few mottles are observed. Numerous non-vegetated profiles exhibit very dark brown to black stained surface layers characterized by a high organic matter content. This morphology is attributable to a previously existing surface horizon that has been removed by silt and sand sized material.

Brunisols

The Brunisols occur in well to moderately well drained sites, from sea level to 400 m a.s.l. Shallow to deep organic horizons overlie dark brown to black sandy loam mineral horizons and very dark brown to yellowish brown sandy loam to sandy clay loam subsurface horizons. Field textures are consistently more clayey than laboratory analysis suggest. Upper mineral horizons have weakly developed subangular structure which may persist to depth. Bh horizons occur in many profiles and in some the B horizon is very compact. Yet other profiles exhibit yellowish brown subsoils with convuluted iron pan horizons (Plate 8.1). C horizons may be thin to relatively deep and mottled.

Soils with similar characteristics to some profiles have been described in upland areas in Scotland (Romans & Robertson 1974), in the Orkney and Shetland Islands (Dry, Scottish Soil Survey, personal communication). On some upland plateau areas, over frost shattered bedrock and undifferentiated till some Brunisols have gravelly silt loam surface horizons with stone polygons and soil stripes (Plate 8.2) overlying dark greyish brown horizons which become yellow brown with depth (Husahagi type). These appear to be similar to Arctic Brown soils of well drained periglacial areas (Tedrow 1977).

The clay content of these soils is generally insignificant with sand constituting more than 50 % and silt about 30 %. Loss on ignition in the mineral horizons is commonly greater than 25 %, and they have somewhat acidic reactions. CEC range from 24 to 50 meq and the base saturation from 6 to 40 %. Sodium may dominate the exchange bases, whereas calcium and magnesium together are most common. Illuviation of Fe, Al, and Mn into the B horizons is not consistently apparent in all profiles (Table 8.3) but in soils with arctic polygon formation very clear increases in extractable Fe, Al and Mn were recorded with depth.

The Husahagi type of Brunisol supports stone polygons and soil stripes on slopes less than 4° and as elongated sorted nets and sorted stone lines on slopes from 5° to 12°. An exhumed soil polygon (Plate 8.3) reveals relatively fine homogeneous textured material with distinct dark gray separations. The depth of solum is generally less than 55 cm, textures are a gravelly silty loam and a weak granular structure in the surface horizon grades into a moderate to medium subangular structure. Gleying, mottling, and a marked accumulation of organic matter in the lower 20 cm are often encountered, due to the relatively impermeable nature of the bedrock parent material.

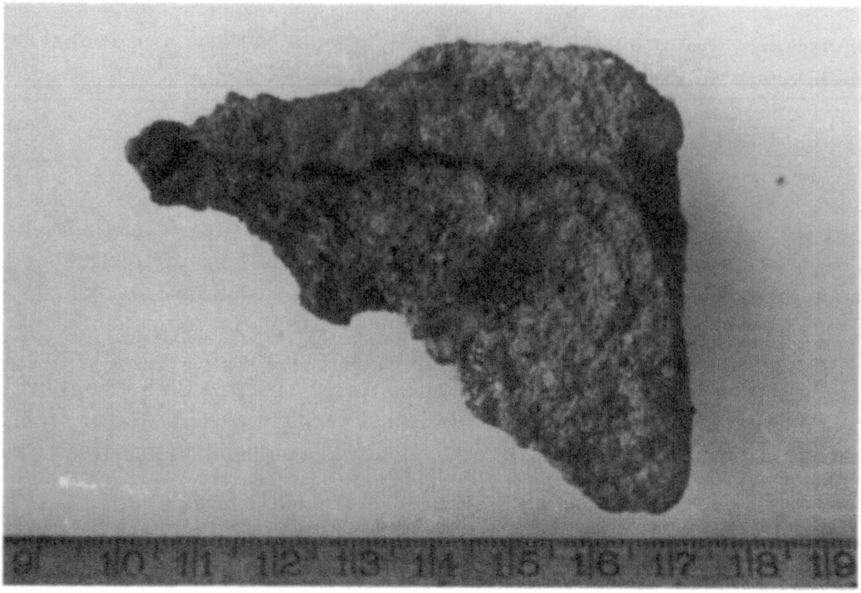

Plate 8.1. Macrophotograph of a strongly developed thin iron pan.

Plate 8.2. Stone stripes developed on undifferentiated till and frost shattered bedrock.

Plate 8.3. An exposed frost polygon (Husahagi type).

Podzols

The Podzols have thick organic surface layers which are well humified and contain fine sand grains. This organic layer is capable of holding water for long periods, so that the conditions below these layers are often anaerobic.

The distribution and differentiation of the Podzols in the Faeroes are determined by variations in the texture of the parent material, altitude, form and angle of the slope, site stability and depth to an impermeable pedologic horizon or bedrock. *Festuca vivipara*, F. *rupra*, *Calluna vulgaris*, *Nardus stricta* and mosses form the dominant vegetation complex.

Podzols have thick organic layers evenly overlaying dark brown sandy to silt loam mineral horizons. The presence of a thin, often strongly convoluted ironpan immediately under the upper mineral horizons is characteristic of these

Table 8.3. Contents of extractable Fe, Al and Mn in some soils of the Faeroe Islands.*

Horizon	Depth cm	Fcd %	Fco %	Fep %	Ald %	Alo %	Alp %	Mnd ppm	Mno ppm
Regosolic soils									
Om	8–0	1.92	1.00	1.20	1.56	2.58		160	68
A	0–19	2.08	1.08	0.68	1.46	3.18		140	42
C1h	19–32	3.52	2.18	2.05	2.04	2.68		69	19
C2c	40–40	4.52	0.52	1.56	0.14	3.20		75	80
C2	50–62	1.00	0.18	0.31	0.46	1.28		45	
Brunisolic soils									
Oh	4–0	1.68	1.36	1.04	1.32	2.62		55	110
Al	0–20	1.78	1.62	1.70	1.66	2.40		280	165
	20–43	2.80	1.82	1.38	1.40	1.22			
Bh	43–75	2.22	1.83	1.80	1.39	1.62		90	180
	75–95	2.00	1.50	1.02	1.24	1.50		80	195
	95–108	0.94	0.85	0.84	1.08	2.32			20
Podzolic soils									
Al	0–20	1.10	0.76	0.52	0.62	0.94	0.58	34	10
Aej	20–35	1.48	0.96	0.72	0.40	1.00	0.64	35	6
Bli	35–36	0.68	0.40	0.50	0.22	0.82	0.20	35	13
B2c	36–87	0.68	0.42	0.20	0.20	1.20	0.20	35	6
B3c	87–120	4.92	1.70	0.03	0.14	0.62	0.14	120	60
Cc	120–150	0.80	0.20	0.06	0.08	0.42	0.08	80	26
Gleysolic soils									
A	4–0	1.30	0.90	1.00	0.58	0.64			60
Ag	0–17	4.20	2.62	0.76	0.84	0.76			
B1g	17–26	1.20	0.80	0.48	0.64	1.30			11
B2g	26–55	2.20	1.38	1.27	3.32	4.78			
B3g	55–69	0.48	0.34	0.38	2.72	2.06			18
C	69–	1.08	0.40	0.42	0.76	0.78			4
Organic soils									
	0–22	1.62	1.04	1.40	1.44	2.70		120	420
	22–40	2.28	1.80	1.64	1.60	1.75		180	
	40–55	0.98	0.40	0.62	0.38	0.40		690	100
	55–177	0.90	0.41	0.62	0.32	0.56		300	150

*d = dithionite; o = oxalate; p = pyrophosphate.

soils, below which the horizons are brown or yellowish brown. There is a light coloured and weakly to moderately compacted incipient Ae horizon (Plate 8.4) and the B are mottled. Bog iron ore may occur. The Podzols occur on gently sloping upland surfaces on tills and colluvium under 100 m a.s.l. Total solum is usually less than 65 cm.

The Podzol soils are strongly acidic in the upper horizons and commonly become markedly less with depth; pH of 5.5 in a C horizon was the highest measured in the islands. Caly contents are less than 5 % and sand usually varies from 60 to 80 % although in several profiles the deeper horizons contained substantially more silt than sand. The organic matter content of the mineral horizons is the lowest for any group in the Faeroes. Values for Fed, Feo, Fep, Ald and Alo are lower than for most other mineral soils, but they increase sharply below the ironpan horizon (Table 8.3).

Great soil group	
Canada	U.S.D.A. (1975)
Regosol	Cryorthent
Brunisol	Mainly Fragiorthod, some Udipsamment
Podzol	Placaquept
	Fragihumod
Gleysol	Fragiaquept
Organic Soils	Medihemist

Plate 8.4. Strongly convoluted, thin iron pan developed in a Podzol. Parent material is dense, sub-angular basal till.

Gleysols

Gleysols display a variety of morphologies but usually occur as gray, medium to fine textured surface mineral horizons high in organic matter grading into gleyed, brown, loamy sand subsoils. Organic staining and manganiferrous concretions may occur in the B2g horizons. Humic surfaces and peaty layers commonly overly gleyed mineral horizons. There may be a dense, black, organic layer at the bedrock contact in shallow examples of this group. They usually occur in poorly drained sites on backslopes or cirques and low lying regions adjacent to organic soils up to about 400 m a.s.l.

The clay content is higher in these soils than any other groups ranging from 5 to 8 % and although organic matter is relatively low it increases significantly in the B2g horizon. Exchangeable Mg is quite high in surface horizons (3.5 meq/100 g) followed by high values for Na and Ca. Surface inundation by runoff from higher slopes may account for the cation illuviation in the A horizon. Extractable Fe and Al values in the B2g are significantly lower than those in the horizons above and below.

Organic soils

Soils in the Faeroes having organic surface layers of 40 cm or more in thickness, with at least 30 % organic matter have been classed as Organic soils. These soils range from weakly decomposed peats (H3) on the von Post scale (1926) to well decomposed peats (H6) and have high amounts of fibric and mesic materials. The mineral fraction is erratically distributed through these soils and is manifest as lenses and sedimentary layers or merely dispersed throughout the organic material. The size composition is commonly about 30 % fine sand, 50 % silt and 20 % clay-sized materials. The organic horizons overlie gray, gravelly, loamy sand subsoils and are waterlogged for most of the year. This soil group is found from sea level to over 700 m a.s.l.

These are the most acid soils in the islands with the highest cation exchange capacities. Base saturation is less than 10 % in the surface and becomes greater than 20 % with depth. Sodium is the highest exchange base with calcium significantly higher than magnesium. Extractable Fe and Al are significantly higher in the surface than the deep horizons whereas extractable Mn is very high and indeed the highest value measured in the Faeroes. However, in some profiles the extractable Mn values were the least measured in the island.

The alpine peat or Tjornuvik type form in shallow depressions and rock knob lakes. After intial deposition of fine alluvial and aeolian material in bedrock catchment areas, the soil surface is characterized by having a 2 to 4 cm surface layer of very fine silt material. Generally, this surface remains moist but upon drying exhibits a marked polygonal cracking pattern. This moist surface serves as an effective medium to encourage the entrappment of local aeolian (organic and inorganic) material. These surfaces are then colonized by Wooly Hair moss which due to the high humidity and lower temperatures produce an accumulation of plant residues. The dominant vegetation is *Carex goodenoughi* and various other sedges. The profile morphology presents a nearly continuous, thin layer of mosses or coarse grasses abruptly overlying the water saturated alluvial and aeolian parent materials. Horizon differentiation is weak despite a tendency to develop alternating layers of organic-mineral material to a depth of 50 cm. Dark gray, black and dark brown colours throughout the profile are acquired from the decomposition of organic matter and some minor oxidation-reduction processes operating as a result of a fluctuating water table. These soils have 60 to 70 % mesic and from 30 to 40% fibric organic material.

Organic-peat soils, the Saksen type, dominate areas with dense, heterogeneous ground moraine and steepland till. These areas often have a marked hummocky topography and occur along the slopes adjacent to the cirque floors and the U-shaped valleys of the Faeroes. The dominant vegetation is *Carex goodenoughi*, C. rigida, *Calluna vulgaris* and mosses. These soils have a black open structured mesic organic material in the upper surface horizons. Below a depth of 60 cm, the organic matter is well decomposed and has a massive structure. The peat material is very sticky when moist and hard upon drying.

Plate 8.5. A thin, moist 2–6 cm layer of silt sized material and gravels surrounded by grasses serve as an effective medium for the entrappment of aeolian material.

Peat deposits of the Faeroes have been reported as being relatively thin throughout the history of the Faeroese civilization (O'Dell 1957). Although once a very important energy source and social activity (Williamson 1948), the contemporary large scale cutting of peat is now confined to a few localized lowland areas (Saksendalur, Streymoy; Sandur, Sandoy). Numerous coarse sand and gravel layers are evident throughout the subsoil horizons and may indicate inundation of the profile by sheetwash materials from upslope. The Saksen Organic soil represents an intermediate type of peat material as it is fed by precipitation and is regularly inundated by nutrient rich groundwaters. A mor type humus develops and accumulates to form a permanently waterlogged and anaerobic horizon. Under these conditions further decomposition of plant material is inhibited and the peat layer begins to grow upwards (Russell 1971). This layer is relatively impermeable and a rise in the watertable accompanies the upward growth of the deposit. Large continuous deposits of basin peats occur throughout the Faeroe's landscape.

Accumulations of wind borne material have and still strongly influence organic soil formation giving rise to complex profile morphology. Violent winds destroy and rip up carpets of vegetation from the fragile upland soils. These areas then provide a source areas from which silt-sized material may be removed. This wind borne material is then captured in wet areas along valley floors and backslopes and vegetation begins to grow (Plate 8.5). Subsequent entrapment of wind blown organic and mineral matter, redeposition of alluvium and peat formation produces a profile characterized by alternating layers of mesic peat development and fluvially deposited medium textured sands and gravels.

Soil/landscape relationships

Although the Faeroe Islands occupy a small area, they are rugged and mountainous so the soils exhibit a moderate range of differentiation. At altitudes above 400 m, the landscape is usually steep and strongly exposed to wind and water erosion. Regosols are by far the most common soil, although some Brunisols occur in better drained sites on cirque floors and Organic soils, in the poorly drained plateau areas of considerable extent, occur between 300 m and 400 m. In the flat and gently rolling areas between these altitudes Brunisols, both with and without various forms of patterned ground, are most common. Gleysols and Organic soils occur in the depressions and Regosols on the steeper slopes.

In the glaciated valleys and valley floors below 300 m, Podzols occur on gentle slopes, Brunisols in similar but less well drained sites, Gleysols in minor depressions, Organic soils in basins and Regosols on the steep slopes.

Soil micromorphology

The micromorphology of the Faeroe Island soils is typical of soils with modest pedogenetic development. Soil fabrics comprise subangular skeletal grains of very uneven sized distribution, randomly set in a somewhat homogeneous s-matrix. They have mainly porphyroskelic related distribution patterns with minor agglomeroplasmic in some Brunisols. Most thin sections had isotic plasmic fabrics, silasepic in some B horizons and minor skelinsepic in an eluviated Podzol horizon.

The soils are characterized by high amounts of void space consisting of an intricate system of interconnected and mammillated vughs and channels. Organic matter commonly occurs as black to reddish brown nodules around skeleton grains in plain light. In Podzolic Ae and Aej horizons, fine pellety amorphous organic matter loosely binds mineral grains. No true cutans were observed although in Podzols subcutanic features resembling diffusion cutans and composed of sesquioxides, manganese or organic matter occur throughout the solum.

Pedotubules are not a common pedologic feature but a few occur predominately as ortho-isotubules or ganotubules. Orthic glaebules form commonly in the Faeroe Islands soils due to the in situ accretion of the soluble sesquioxide and manganous constituents. These nodules are usually diffuse and are highly irregular to prolate indicative of the process of gleization. Glaebules of iron oxide occur frequently in the s-matrix around old root channels of the poorly drained soils. Glaebular haloes composed of the soluble organic and sesquioxide materials are frequent and emphasize the in situ formation of the nodules by accretion.

The pedological organization of the Faeroe Island soils is typical of soils with weak to intermediate development. Orthic pedological features are uncommon, plasma concentrations although frequent are diffuse and irregular and plasma separations are almost non-existent. Thin sections of mineral soils are dominated by skeleton grains.

Clay mineralogy

Clay mineralogy was determined by x-ray diffraction and the quantitative determinations by computer programming. It appears that the parent rocks have been gently but thoroughly metamorphosed to a zeolite grade of metamorphism so that they are composed of pyroxenes, labradorite, zeolites and some reasonably pure montmorillonite with some interstratification of chlorite with montmorillonite with some interstratification of chlorite with montmorillonite. Subaerial weathering produces chlorite very quickly so that orthochlorite may be considered as an inherited mineral in the true soil parent material. Modest evidence suggests that with increased weathering, such as in Podzol pedogenesis, chlorites tend to form likely at the expense of montmorillonite. This is not consistent with prevailing concepts on mineral formation in Podzolic soils, wherein montmorillonite is considered to have formed from chlorite.

No information is available on the paleostratigraphy of the till and other possible soil parent materials, so the origin of illite may be merely conjectured: at one low-lying site it may be in materials longer exposed to weathering than generally current in the archipelago. Similarly, the presence of kaolin is likely inherited from rock materials. The lack of olivines in silt and clay-sized materials suggest that this mineral is unstable in such fractions, although no specific minimum size to its physical stability may be intimated from these studies.

Authors' addresses:
Dr G. K. Rutherford
Dept. of Geography, Queen's University,
Kingston, Ontario,
Canada

Mr C. E. B. Taylor
Ontario Hydro,
Toronto, Ontario,
Canada

9. Man and the physical environment

J. P. Joensen

The first account of the Faeroe Islands was written by the Irish monk 'Dicuil' about 825, in which he gives an account of, amongst other things, the Faeroe Islands as being a place where some Irish monks lived but were later ousted by Nordic Vikings. In addition, he indicated that there were many birds and sheep on the islands.

It is believed that the Vikings began to inhabit the Faeroes about 800 A.D. Pollen analysis has given an indication of an earlier settlement, but this has not been confirmed by archeological evidence. Culturally, the Faeroes are Nordic, but there is some doubt as to whether the settlement of the Faeroes occured before or after that of the other Atlantic islands, where the Vikings had also settled. There are certain elements in the culture, together with some physical anthropological characteristics, which suggests a Celtic connection. At present, there is no solution to this question.

Relics of Faeroese culture and vocation are preserved in many ways. The unique expression of Faeroese culture can be partly explained by geographic isolation, in that they have not been influenced by the cultural changes and innovations which took place in the other Nordic cultural areas.

In the period between the original settlement and up until 1500, there was a relatively lively connection between the Faeroes and the world at large. The Faeroese had their own ships, with which they sailed to neighbouring countries to trade and collect provisions for the islands. The 12th and 13th centuries were periods of economic advance, some of the evidence being a cathedral which, however, was never completed because of an economic and accompanying cultural decline. This began in the 14th and continued until the 15th century. The cause of this economic decline is thought to be connected with climatic change.

The period of depression was the cause of the Faeroese inability to maintain their own shipping connections with the outside world. Their main trade connections had been with the Bergen in Norway. From then onwards, it was only foreigners who maintained any ship connection with the Faeroes and foreign ships came with provisions a couple of times each summer. In this way,

Monographiae Biologicae, Vol. 46 ed. by G. K. Rutherford

the cultural connections, along with the possibilities for cultural change and advancement, were considerably reduced. The Faeroe Islands and culture were as good as left to their own devices, developing only with considerably limited external impulses.

The civil servants were Danish, and with the exception of a few priests, all lived in Torshavn. Naturally, they had their own lifestyle, but the social distinction between them and the common people was such that their influence would hardly filter down to other levels of society. The other connection to the outside world was trade with other countries. The legal trade had its only outlet in Torshavn until the beginning of the 19th century, after which three centres were established on other islands. Until 1619 all trade was directed through Bergen, but after this it was transferred to Copenhagen, which became their only contact with the outside world.

In the beginning, it was only the private trading companies that received monopolies to trade with the Faeroes. However, they often lacked the necessary interest in the Faeroes' welfare and well-being, so that complaints were commonly raised over the provision of goods. This was particularly bad during the so-called 'Gabel period' (1655–1709), when both father and son Gabel held a complete trade monopoly on the Faeroes. Therefore, in 1709, the trade was taken over by the Danish royal government. There was a royal monopoly on all trade with the Faeroes until 1856, but at the end of the 18th and the beginning of the 19th century, some marked changes in Faeroese society began to take place. These eventually led to a change in their society from a passive and conservative farming community to a modern fishing and industrial society.

As a link in this development, the trade monopoly was abolished in 1856 and free trade was introduced. In the beginning, it was mostly foreigners that established commercial and merchant houses on the islands, some of which still exist. In later years, some of these commercial houses also played an important role in the financing of the fishing fleet. Free trade was also accompanied by the sale of alcohol, which increased enormously. This was not the only expression of the merchants' enthusiasm, as many farmers in this period mortgaged their farms for alcohol. This alcohol consumption was also a sign of a society undergoing change. The old economic social and cultural values were on the decline, whilst new values had not yet been moulded.

By the year 1800 the Faeroes were a peasant society, where practically all economic and social rights were associated with the farm. The land not only gave access to the traditional agriculture — cattle and animal raising of the Faeroes — but certain other privileges were also connected with the family farm (for example, the practice of fowling and the receipt of certain percentages of the total catch). This privilege also extended to the catch of the pilot whale. Farmers had the main power in the community in association with all of these privileges; in particular, the so-called King's farmers.

During the Middle Ages, the church, the king and foreign nobles acquired most of the farming land which was then leased to tenant farmers. There was, however, some allodial (odel) and privately held land belonging to farmers. The Catholic Church was the largest landowner, but after the Reformation church lands were taken over by the king and became the so-called royal domains.

Land owned by the nobility was gradually sold into private hands. In contradistinction to farms in the royal domains, which were kept in indivisible units, the allodial lands could be divided by inheritance and were subject to purchase and sale. The result of this was that, in the Middle Ages, these lands had already been greatly reduced to small sections, while farms on the royal domains were kept at economically-sized units down through the generations. Even though the allodial lands could be collected by purchase, they would likely be redivided in the next generation. Farmers on the royal domains had the most favourable opportunities to accumulate capital for the purchase of allodial land for their young children, who had no other opportunity than as allodial farmers' sons. In this community, strategic marriages were of great importance.

All of the settlement was collected in 'bygdir'. The Faeroese 'bygd' can be defined as a local community with associated pasture and grazing lands — 85 bygdir existed from earlier days. Originally a bygd consisted of a single farm which embraced a large area, but it was not long before secondary economically dependent farms, associated with the main farm, arose. However, even during the Viking times, these had become independent economic units, so that new bygdir originated. Later, these were further divided into smaller units, so that a bygd came to consist of two or more groups of farms; these were called bylingar or in the singular, bylingur (Plate 9.1).

By 1800 most bygdir consisted of several bylingar, and each bylingur was comprised of a collection of one or more farms. A unit farm is called in Faeroese 'eini hús' (plural). The unit farm was comprised of many different houses, each with its own use. The old traditional Faeroese farm house has its cultural historic counterparts in the greater northwest European area, and probably came into being immediately after the longhouse of the Viking period.

This house consists of an internal frame construction of wood with stone outer walls. Originally, one or more external walls were panelled. The roof is raised at about 35 % and consists of grass turf on a birch-bark underlayer, which in turn rests on cross boards lying in the long direction of the house. The main rooms in the house are the hearth room and the window room; many houses were comprised only of these two. Poorer homes only consisted of a hearth room.

The hearth room was a work place and communal room, whilst the window room was more elegant and, as the name suggests, was provided with glass windows. In the hearth room the daylight penetrated only through the door and the hole in the roof (where the smoke escaped) which was usually placed on the end wall. The floor of the hearth room was composed of stamped clay, while in the window room it was covered with floorboards. Above the window room

Plate 9.1. Settlement on the Faeroes occurs in local communities, comprised of a number of farms. (Photo: Kalmar and Alan.)

Plate 9.2. Only a modest portion of land on the Faeroes is cultivated, primarily grass, grain and potatoes. (Photo: Kalmar and Alan.)

there was an attic, whilst the ceiling in the hearth room was open. Oil lamps were used for light.

At the end of the 19th century, many changes in the Faeroese building customs took place, giving rise to another kind of house. These, which are increasing in number, consist of a wooden house with a ground floor and attic, and are raised on a stone foundation which constitutes a cellar. In earlier times, the cellar was used for animal stalls. In the traditional house the stall was in one end, but the animals and people shared the same entrance. In many instances, the stall was a separate building. The other outhouses lay at certain distances from the main house. Most important of these was the store house, which was called hjallur, with large farms having several of these. Aside from food, one also kept clothes in these houses. There are many different kinds of hjallur, the most common being a rectangular house of the same frame construction as the main farm house. The walls and gables are clad with vertical slats at a distance of about 1 cm from each other. The gables, however, are made of stone. In the northern islands, one finds hjallurs which are made of half stone with the other half covered with vertical slats. A common feature to all hjallurs is that they are built to allow the wind to blow readily through them. The same principles were involved in stone houses which were used to conserve turf, as this was the only fuel they had apart from drift timber.

No forests grow on the Faeroes. Grain does not usually ripen, therefore a special grain drying house is necessary where one can dry corn by burning turf. This house, which was built of the conventional stone with a grass turf roof, was usually placed at a reasonable distance from the farm proper because of fire danger. Both in appearance, form and colour, the old architectural style was well blended with its surroundings.

Cultivated fields in Faeroese are called bøur'. Uncultivated but improved pastures are called 'hagi'. Bøur and hagi are sharply divided with stone and other forms of fencing. There is only a modest part of the Faeroese land which is cultivated-about 7 %. Cultivated fields are used for grass, grain and recently, for potatoes (Plate 9.2). The Faeroe Islands lie in a climatically marginal zone for agriculture, so that grain ripens poorly. Six-row barley is the only grain that has been grown. Grain and grass constitute a simple rotation. Each seventh year the field would be sewn with barley and in the ensuing years, grass would be grown. Grass growth would be luxuriant for several years after.

In former times, ploughs were practically unknown. As an implement, two forms of an unique Faeroese spade (haki) were used; one to cultivate and the other for collecting turf for fuel. It is common to see part of the arable land built up into terrace-like formations called 'teigalendi', a particular form of cultivation whereby the furrow, after having been cut by the haki, is inverted and laid adjacent to the previous one. This method of cultivation is called 'reinavelting' and was almost entirely used for grain. When grain cultivation declined at the beginning of this century, this method went out of use.

Aside from indicating pastures, the hagi is also an administrative unit; each hagi consists of a certain number of merkur, which are further divided into gyllin. The cultivated land is divided into discrete areas between the owners within a hagi, in relationship to how many merkur or gyllin there are. However, the pasture lands are a kind of community area in which one has rights according to the size of one's property. Each hagi constitutes a defined area for a certain number of sheep, which pasture outside all year around. The utilization of the hagi is a function of a simple ecological balance where flocks of sheep form the link between nature and culture (Plate 9.3).

In the winter the arable land is used for pasture, as in the summer one attempts to get the sheep grazing in the mountain areas, which can not be used in winter due to frost and snow. This must be carried out as part of the normal grazing pattern. In a hagi, the sheep have been trained to keep within well-defined areas, and flocks of sheep acquire names after their areas. It is of special importance that the size of each flock is kept in proportion to its grazing area. If the flock is too large, then there is a temptation to force it in on another flock's area. If, on the other hand, it is too little, other sheep can gain access to the area. For many years, the Faeroese made maximum use of the pasture areas in this way, without having to invest in fences and similar equipment. Training the sheep to keep themselves in certain areas is also important, particularly when they are brought into the fold several times per year. The sheep are shorn in summer, and in the autumn some are taken out for slaughter.

As far back as one knows, two forms of property rights have been used on the Faeroes. One form is where each sheep is owned by a certain individual as private property who may bear either gains or losses in flock numbers. This form is called 'kenning'. The other form is when the sheep farmers own all the sheep in the community in common so that any gains or losses are divided equally. This form is called 'felag'. The sheep on the kenning are, as a rule, tamer than those on the felag. The principles used in sheep breeding are the same as in the 1200's as are the slaughter methods. In the farming society, wool and wool products were the Faeroes' most important export item. Mutton, then and now, is completely consumed within the Faeroese community. Meat may be preserved by wind drying alone.

Fowling rights fell to the landholder, and were an important part of community life in the parishes which had nesting cliffs nearby. There are some rather clear rules stating how much of the catch falls to the land owner. Usually this is about $\frac{1}{4}$ to $\frac{1}{2}$ of the total catch (when over thirty birds); the remainder falls to the fowler. The part which is due to the farmers is divided according to the number of landowners.

Earlier, wild fowl constituted an important part of the diet. The feathers were used for comforters and pillows. The birds that were caught were similar to birds caught today, with the important exception of a new element — the arrival of the Storm Petrel to the Faeroe Islands at the end of the last century. In

Plate 9.3. Sheep serve as a link between nature and culture, grazing in mountainous areas in summer and arable land in winter. The sheep are shorn in the spring and slaughtered in the fall. (Photo: Kalmar and Alan.)

distinction to the other birds which are hunted in summer, this one is hunted in winter.

When one excludes the Puffin, which can hatch on relatively gentle slopes, most birds, including the Puffin and the Guillemot, hatch on steep, isolated and inaccessable nesting cliffs. In several places the fowlers must be let down on ropes up to 100 m long. Puffins are usually caught at the beginning of the season, whilst they are sitting in their small caves. They are seized either in their bare hands or with the aid of a hook, but they can also be taken in the same way as the Guillemot — with a net. The net is spread between two $1\frac{1}{2}$ m long wooden poles, which are attached to the end of a 4 m long stick. There are two forms; one has a short stick and rather long net called 'fyglingarstong' that was used to hunt Guillemots, whilst they sat in a nitch on a nesting cliff. The other net is shorter but on a longer and more flexible rod called 'fleygastong' which is still used to capture flying birds. Fyglingarstong is probably the oldest form, and is not used any more, being last used on Skúvoy in the 1920's. On Skúvoy, until a few years ago, Guillemots had been caught in the nesting cliffs, where the egg collectors were let down on a long line which was so heavy that most of the island's inhabitants were required to haul it up again (Plate 9.4).

Skúvoy and Mykines are the most important bird islands; Skúvoy is the island for Guillemots while Mykines is the island for Puffins, and where Gannets have their only hatching ground. A hunting net is not used, but the eggs are taken directly with the bare hands from the nitches in the cliff. The old birds are caught at night by the hunters, who are let down over the cliffs then lie themselves completely over the sleeping birds and ring their necks. The birds, which are almost ready to fly, are knocked down in daylight with a club called 'kádissa'. Up to ten years ago, fowling was practiced under the old traditions involving social organisations and rituals, but these have now disappeared.

The pilot whale hunt is the most important picturesque part of the hunting life on the Faeroes, and it has not changed for several hundred years. Supporting the pilot whale hunt is an entire social institution, with its organization, commando structure and ethics, as well as rules for the division of the catch. There are also regulations as to how the siting of a school of whales is propagated, and where the whales are to be driven. These regulations came into force through an act of parliament in 1832 (Plate 9.5).

Whales live in schools, whose size usually ranges from 50–500, though it can be larger or smaller. As soon as the school is detected, fishing boats and other vessels collect around it and the chase begins. The whales are chased into shallow waters at the end of narrow bays and fjords. There are a limited number of bays and fjords where whales are allowed to be killed. Slaughtering is done by using a special knife (grindaknívur) and a whale spear (hvalvákn). Wherever possible, slaughtering is done by knife alone. A deep hole is cut in the whale's neck, a hands length from the blowhole, so that the spinal cord can be cut. This is accomplished either by wading out to the whale on a flat beach or by

Plate 9.4. On Skuvoy, the eggs of the Guillemot are gathered on the nesting cliffs. (Photo: Kalmar and Alan.)

approaching close to the whale in a boat so that the necessary cut may be made. In deeper water, the whales are sometimes killed with whale spears. It is part of the whaling tradition that the killing be carried out as quickly and efficiently as possible. Even in our technological age, it has not been possible to find a more effective method to kill whales.

The meat and blubber have been of invaluable importance to the farming community. Calculations indicate that, over a long period of time, whale meat and blubber have constituted about a quarter of the daily calorie requirement. However, pilot whales erratic occurances have caused this source of calories to be somewhat unreliable. In some years, several hundred whales may be taken and in other years, only a few.

In earlier times a large part of the catch was considered to be the property of the farmers- originally a half, later a quarter. In the early part of this century this privilege was abolished. Now the whale meat is divided according to the following principle — all the residents in a pilot whale district get the same portion of the catch, regardless of whether they participated in the catch or not. However, in later years, it has become practice that participants in the chase itself receive something extra. From a purely economic point of view, the pilot whale chase is probably unnecessary, but it is part of the Faeroese culture and the pilot whale and blubber are an important and highly prized part of the diet.

Seal hunting never attained the same dimensions as whaling because seals can not be caught in large numbers. They were hunted in dark caves where they have their young, or on skerries along the coast, and were knocked down with a club (kópagassi). The hunting of large whales has now ceased on the Faeroes. It was introduced by Norwegians about 1900 when they established several whaling stations. Several of these were taken over by Faeroese companies.

For the farming community, fishing was always a part of the natural householding and some of the stock fish was exported. However, wool and wool products have been of considerably greater importance. In the daily diet, the grey coalfish was of great importance and could be caught in such large amounts that, when dried, it could be used as animal food. They were fished by rod close to shore, and many were taken in a dragnet. Cod and similar fish were taken in nets in deeper waters, at various distances from the coast. The fishing tackle was very primitive — a hook, with a stone as a sinker, attached to a line. The fishery was only one of the many economic ways of life for the farming community. However, this later provided the economic background for a highly industrialized fishing community.

For the farmer, there were many factors which hindered the development of fishing. This was partly because fishing was a part of the work of the farm. Farmers had an obligation to keep a boat, and had a right to a crew. This could, in an extreme case, allow a farmer to stop someone who wished to fish his own boat. At the end of the 18th century, a number of people began to use fishing as their main livelihood and this number began to increase, so that during the early part of the 19th century, the structure of the community slowly began to change. The economic foundation was changing; wool and wool products fell in price, whilst fish prices remained stable.

A new progressive kind of civil servant arose. The Danish governor of the Faeroes and several other Faeroese, took a study trip in 1839 to the Shetland Islands, in order to observe how the Shetlanders produced dried cod and salted fish. Shortly afterward, the royal trade monopoly began to buy large amounts of fresh fish. Several attempts were made to introduce Danish covered fishing boats, but the Faeroese were not used to these, so the attempt was unsuccessful. In 1810, however, the Shetlanders began to fish cod on the Faeroese banks, using larger sailing ships called 'smacks'. They often came into fjords and bays

Plate 9.5. The pilot whale hunt is a very picturesque ritual of the Faeroese culture. It has been regulated by Parliament since 1832. (Photo: Kalmar and Alan.)

Plate 9.6. the Faeroese gained experience in fishing from the Shetlanders and after 1900, the ships rapidly increased in number to over 100. (Photo: Føroya Fornminnissavn.)

where the Faeroese got to know them and some even worked on board the ships during the summers. In this way the ordinary Faeroese gained experience in fishing with such ships, so that it was much easier to introduce ocean going fishing vessels. In 1872 some men from Tórshavn bought a sloop from England, introducing a new period of economic independence in fishing. Before 1900, there were only a few ships, but this number increased rapidly to a fishing fleet of over 100 ships. This did not mean a cessation of agricultural production and activities, for they remained an important part of the natural householding (Plate 9.6).

The population has increased a great deal since 1800, when it was about 5300. Around 1850 the population had increased to about 8000 and to over 15 000 by 1900. Today there are more than 40 000 inhabitants. The importance of the farmers dropped, in favour of shipping owners and captains, who were becoming an influential part of the community. This gave a new impetus to those who did not own land and could only work on a farm. They could now get employment as fishermen and earn sufficient money to build their own houses and have their own families.

Sloop fishing consolidated its position, and continued more or less unchanged to the second World War. In these forty years the annual rhythm was the same — fishing began early in March and was completed at the end of September when ships were laid up for the winter. Even though most of the men were out fising for the entire summer, there was a sufficient work force left behind to manage the necessary agricultural work, such as haying. The fishermen could carry out part of the prepartory work in the spring before they left, and some during the time between fishing trips. Finally, they would be home in time to bring in the hay and harvest the potatoes. Sloop fishing was an integral part of many cultural livelihoods, being that part which gave ready cash. The economy of each household depended upon the number of supplementary endeavours it had, or if they had decided to live entirely from the money acquired from sloop fishing. In the winter there was some fishing using small boats.

If we exclude the fish which was taken on ice directly to markets outside of the Faeroes, all other fish (mainly cod) was salted. Most of this was exported as such, but some was dried in the sun and exported as dried fish. Fish drying created a completely new form of work in the summer months, particularly on Suðuroy where many people migrated in the peak season. The process of drying became mechanized, so that the fish was dried indoors using hot air.

The old cultural characteristics of the farming community contined into the 20th century, but with the outbreak of the second World War a new change in culture and community began. In the course of twenty to thirty years, the Faeroes developed into a highly industrialized fishing society where the old hunting and agricultural culture continued as a relic, or as a cultivation hobby.

Prior to the second World War, handlines and longlines were the most important fishing tools and in 1939, there were only ten trawlers. During the war, the Faeroes were occupied by English troops and the new contact which was established brought new opportunity to the fishing industry. England took large quantities of fresh fish, which was transported on ice. This continued in 1946, but the usual method of salting fish had also begun. After the war, many used trawlers purchased from England, which did not always prove to be economical, but from the end of the 1950's a systemic renewal of the fishing fleet began. There has been a particular concentration on new ships, which has continued to produce one of the world's most modern and versatile fishing fleets.

Not only the fishing techniques have become more varied, but the products have also become more differentiated. The modern fishing vessels are floating factories, which deliver their products packed and ready to cook. There has been an increase of fishing in home waters over the last ten years, which has made possible a considerable investment in the fishing industry since the middle of the 1960's. Fishing in distant waters by Iceland, Greenland, Newfoundland and in the Barents Sea have previously been dominant, but now fishing near home waters is greatly increasing (Plate 9.7).

Exports from the Faeroes are almost comprised of fish and fish products, so that the fishing industry accounts for about 30 % of the gross national income. In comparison, the agriculture constitutes only 1.8 %. Before the second World War, the Mediterranean countries and England were the most important importers of Faeroese fish products, but Brazil also became important after the war. A great deal is still exported to Mediterranean countries and England, but since the 1960's the market has widened to embrace several countries. The United States and Denmark now top the export list. By far, most of the imports to the Faeroes come from Denmark and other Nordic countries. The majority of the imports are consumer goods, followed by machinery, instruments and industrial transport vehicles; approximately the same amount in value in raw materials are also imported.

Generally speaking, the period since the second World War has been one of economic advance, if one excludes some instability during the 1950's. There has been sufficient profit from commercial activity to allow for considerable investments from both private and public institutions. Most of the private investments have been towards means of production, whilst public investment has included large sums towards harbours, roads, tunnels and bridges, as well as a considerable amount to local coastal traffic. The favourable road network has led to a car density of one per four persons. In addition to ship traffic abroad, there have been regular air connections since the 1960's. All passenger traffic used to go over Denmark, but several years ago regular Faeroese connections were established to Scotland, Norway and Iceland. The housing sector has expanded strongly with high building standards; the object of housing is for each family to live in its own house.

Since the war, there has also been great investment in culture and education. A school of navigation, high schools, a technical college, and community

Plate 9.7. Modern fishing vessels, commonly of local construction, serve as floating factories and often venture into distant waters. (Photo Kalmar and Alan.)

college have been built. In 1965 a Faeroe academy was started. A radio station was established in 1957, but as of yet a television station has not followed this development. A couple of private television associations received temporary permits two years ago to tansmit television programs to small localities. The programs sent from these are usually copies of Danish television programs. The Faeroe administration has, however, plans to establish a Faeroese television station in the immediate future.

The Faeroese have their own language, which is west Nordic, but Danish is learnt very thoroughly as well as other foreign languages. The geographic situation of the Faeroe Islands, with their cultural isolation through many centuries, has preserved many cultural characteristics which have disappeared in the other Nordic countries; the ring dance, songs and ballads from the Middle Ages are a few examples. The changes in society occuring at the end of the last century provoked a national consciousness and consideration for old cultural values, contemporaneously with the development of new artistic expression in word and picture. Painting is perhaps the most important art expression on the Faeroes, but there are also many fiction authors.

Politically, the Faeroes are an important part of the Danish kingdom. They have the same currency, but otherwise the administration is divided so that the Faeroes are responsible for domestic affairs, the Danish state for others and some are administered collectively. The Faeroes received home rule in 1948, and before that, they were a Danish province. The legislative powers, according to the Home Rule Act, rest both with the Danish parliament and with the local Faeroese parliament, or 'Løgting'. There are some affairs which require action from both sources. The local Faeroese government is called 'Landsstyrið. The Faeroes have common foreign policy with Denmark, but are not members of the European community.

Author's address:
Mr J. P. Joensen
Føroya Natturugripasavn
Torshavn,
The Faeroe Islands

Appendix

Glossary of common Faeroese terminology for geomorphic features

Faeroese term	*English equivalent*
Braekker	The talus slope at the base of a hamnar.
Botnur	A cirque; semi-circular valley with steep sides and a flat bottom land.
Bøur	Improved grazing land; infield and outfield that has been cultivated.
Dalur	A valley.
Drangur	Sea stacks; pointed and connected to land as a sea arch.
Eidi	An isthmus or low, flat strip of land, trending east to west that separates two fjords.
Fjallid	A mountain.
Fos	A waterfall.
Gjogv	A more or less deep, steep walled gully created by weathering, streams and marine erosion that hollow out lamellar zones.
Hagi	The open, unimproved grazing land.
Halsur	A broad, long mountain col or saddle; Fangihalsur, Vagar.
Hamnar	Steep head wall, devoid of vegetation.
Lier	The slopes, covered with vegetation, that alternate with hamnar.
Nes	A small promintory of land.
Rudemark	Polygonal ground.
Skard	A narrow mountain col or saddle.
Tangei	A low, eroded headland.
Tindur	A mountain peak.
Ur	Stones and boulders at the base of a hill slope.
Vatn	A lake.
Vik	Small fjord.

Bibliography
to chapters 1 to 5 and 7 to 9

Canadian Soil Survey Committee 1978. The Canadian System of Soil Classification. Can. Dep. Agric. Publ. 1646. Supply and Services Canada, Ottawa. 164 pp.

Currie, J. 1906. The Faroe Islands. The Scottish Geographical Mag. XXII, 61–76, 134–147.

Daly, R. A. 1934. The Changing World of the Ice Age. Yale University Press, New Haven, 273 pp.

Debes, L. J. 1673. Faeroa et Faeroa. Reserte & Co., København.

Forchhammer, J. G. 1824. Om Faeroernes geognostiki Beskaffenhed. Vitenskab Selskabets Skrifter, København.

Geikie, J. 1880. Geology of the Faroe Islands. Trans Roy. Soc. Edin. 30(1): 217–269.

Grossman, K. and J. Lomas. 1895. On the glaciation of the Faroe Islands. The Glacialist's Magazine, June, 1895: 1–15.

Jensen, A. S., Lundbeck, W., Mortensen, J. and R. Sparek (eds.) 1928–1942: Zoology of the Faroes. Vol. 1–3, Copenhagen.

Killerich, A. 1928. The Geography, Hydrology and Climate of the Faroes. In Jensen, A. S., Lundbeck, W., Mortensen, J. and R. Sparek (eds.) The Zoology of the Faroes. 1(1): 1–51. København.

Lewis, C. A. and G. M. Lass. 1965. The Drift Terraces of Slaettaratindur, The Faeroes. Geographical J. 131(2): 247–251.

Noe-Nygaard, C. C. and J. I. Rasmussen. 1957. The making of the basalt plateau of the Faroes. Museum de mineralogue et de geologie de l'universite de Copenhague. Communications Geologiques No. 101, Mexico, D.F.

Ostenfeld, C. H. 1901. The Land Vegetation of the Faroes, in Warming, E. (ed.), Botany of the Faroes based on Danish Investigations. London.

Pennington, W. 1970. The History of British Vegetation. The English Universities Press Ltd., London. 152 pp.

Rasmussen, J. 1963. Recent studies on the Geology of the Faroes. North Atlantic Brota and their History. Oxford University Press: 29–44.

Rasmussen, J. and A. Noe-Nygaard. 1970. Geology of the Faroe Islands: pre-Quaternary. Translated by G. Henderson, Geological Survey of Denmark I, Series No. 25, 139 pp.

Romans, J. C. C. and L. Robertson. 1974. Some aspects of the genesis of alpine and upland soils in the British Isles, in Rutherford, G. K. (ed.), Proc. 4th int. Working-Meeting on Soil Micromorphology. Limestone Press, Kingston, Ontario, Canada: 498–510.

Schrøder, N. 1973. The Faroes — what lies underneath. Welcome to the Faroes — 1973. Anders Nybord A/A, København: 25–32.

Sundt, H. A. 1976. Sheep and Soil in the Faeroe Islands. Examples from an old Suderoy settlement. Norsk geografiske Tidsskr. 30(3): 115–125.

Tarling, D. H. and N. H. Gale. 1968. Isotopic dating and palaeomagnetic polarity in the Faroe Islands. Nature 218: 1043–1044.

Trap, J. P. 1968. Danmark (Faerøerne). Volume XIII, G. E. C. Gads, København, 375 pp.

Trevellyan, W. C. 1835. On the Vegetation and Temperature of the Faeroe Islands. Edin. New Philosoph. J., XVIII: 1–16.

Valentine, K. W. G. 1970. Use of land classification concepts as a basis for a reconnaissance soil survey in Northeastern British Colombia. Can. J. Soil Sci. (1): 71–77.

von Post, L. and E. Granlund. 1926. Sveriges geologiska Undersökning. Ser. C, No. 335, Stockholm.

Ødum, H. 1923. Traek af Faerøernes Morfologi. Forhandlingar vid det 17de skandinaviskare møte, Gøteborg.

Tedrow, J. C. F. 1977. Soils of the Polar Landscapes. Rutgers University Press, New Brunswick, New Jersey. 638 pp.

Index

GPSR Compliance

The European Union's (EU) General Product Safety Regulation (GPSR) is a set of rules that requires consumer products to be safe and our obligations to ensure this.

If you have any concerns about our products, you can contact us on ProductSafety@springernature.com

In case Publisher is established outside the EU, the EU authorized representative is:

Springer Nature Customer Service Center GmbH
Europaplatz 3
69115 Heidelberg, Germany

Batch number: 10607224

Printed by Printforce, the Netherlands